EUTHANASIA IN THE NETHERLANDS

INTERNATIONAL LIBRARY OF ETHICS, LAW, AND THE NEW MEDICINE

VOLUME 20

The titles published in this series are listed at the end of this volume.

EUTHANASIA IN THE NETHERLANDS

The Policy and Practice of Mercy Killing

by

Raphael Cohen-Almagor
University of Haifa,
Haifa, Israel

SPRINGER-SCIENCE+BUSINESS MEDIA, B.V.

A C.I.P. Catalogue record for this book is available from the Library of Congress.

ISBN 978-90-481-6623-7 ISBN 978-1-4020-2251-7 (eBook)
DOI 10.1007/978-1-4020-2251-7

Printed on acid-free paper

In Memory of my thoughtful and learned mentor
Geoffrey Marshall
April 22, 1929 – June 24, 2003
Cherishing his unfailing support
And his infinite wisdom

TABLE OF CONTENTS

ACKNOWLEDGMENTS

This study could not have been carried out without the cooperation of many people to whom I express deep gratitude: Evert van Leeuwen, Arie van der Arend, Arko Oderwald, Ruud ter Meulen, Henk Jochemsen, Henk Leenen, Sjef Gevers, Ron Berghmans, James Kennedy, Bert Thijs, John Griffiths, Jaap Visser, Heleen Dupuis, Gerrit Kimsma, Frank Koerselman, Rob Houtepen, A. van Dantzig, Paul van der Maas, George Beausmans, Egbert Schroten, Johannes van Delden, Chris Rutenfrans, Govert den Hartogh, Henri Wijsbek, Gerrit van der Wal, Bert Keizer and the de B. family. My gratitude is extended also to Bill Winslade, Joyce Appleby, Jo Heth, Diane LeCover and Sharon Woodrow for their instructive comments. I have also benefited from the advice of David Weisstub, and of the anonymous referees of Kluwer, and I thank Zehavit Cohen-Almagor for her help in preparing the Index. The research was greatly enriched by material kindly sent to me by Ejan Mackaay, Randy Wilcox, Derek Humphry, Sylvia Gerhard, Barney Sneiderman, C. van der Meer, Dick Pranger, Henk ten Have and Boudewijn Chabot.

I am also indebted to my research assistants, Dareen and Grace Yaacob, as well as to Martine Bouman, Nurit Litver, the librarians of the Hastings Center, UCLA School of Law, the University of Haifa and Johns Hopkins University for their caring assistance, and to the Fulbright Foundation, the University of Haifa Research Authority, the Israel Academy of Sciences, and the Royal Netherlands Academy of Arts and Sciences for their generous support.

The Institute for Policy Studies at Johns Hopkins provided me with the tranquil conditions which were much needed for the completion of this demanding research project. I am most appreciative of the kind assistance.

Last but not least, I am most thankful to my Dutch friends and colleagues for facilitating this fascinating and thought-provoking research. Their friendship and hospitality made my time in the Netherlands memorable.

Early version of the Preliminaries was published in *New York Int. L. Rev.*, Vol. 16, No. 1 (Winter 2003), pp. 1–42. Part of chapter 2 was published in *Michigan State University-Detroit College of Law Journal of International Law*, Vol. 10 (2001), pp. 319–342. Earlier versions of chapter 4 and chapter 5 were published in *Journal of Law, Medicine & Ethics*, Vol. 30, No. 1 (2002), pp. 95–104, and in *Ethical Perspectives*, Vol. 9, No. 1 (2002), pp. 3–20. Earlier version of chapter 6 was published in *Croatian Journal of Philosophy*, Vol. II, No. 5 (2002); reprinted in *Issues in Law and Medicine*, Vol. 18, No. 3 (Spring 2003), pp. 239–257. Parts of chapter 7 were published in *Formosan Journal of Medical Humanities*, Vol. 3, No. 1–2 (2002), pp. 22–37 and in *Medicine and Law*, Vol. 20, No. 4 (2001), pp. 613–625. Chapter 8 was published in *Theoretical Medicine and Bioethics,* Vol. 23, Nos. 4–5 (2002), pp. 287–303. An earlier version of chapter 9 and parts of chapter 10 were published in *Issues in Law and Medicine*, Vol. 17, No. 1 (summer 2001), pp. 35–68; Vol. 17, No. 2 (Fall 2001), pp. 167–179, and Vol. 18, No. 1 (July 2002), pp. 111–126. Parts of chapter 11 were published in *Medical Law International,* Vol. 5 (2001), pp. 141–159 and in *Synthesis Philosophica*, No. 33, Vol. 17 (2002), pp. 135–155.

The book is dedicated to Geoffrey Marshall, the greatest constitutional theorist I have ever known, and one of the wittiest and sharp personalities I have ever met. Geoffrey was a dear and most wonderful teacher, wise colleague, and kind friend. He enriched my life in many different ways. I miss him greatly and carry his memory at heart.

INTRODUCTION

The Dutch experience has influenced the debate on euthanasia and death with dignity around the globe, especially with regard to whether physician-assisted suicide and euthanasia should be legitimized or legalized. A review of the literature reveals complex and often contradictory views about the Dutch experience. Some claim that the Netherlands offers a model for the world to follow; others believe that the Netherlands represents danger, rather than promise, and that the Dutch experience is the definitive answer regarding why we should *not* make active euthanasia and physician-assisted suicide part of our lives.

Given these contradictory views, it has become clear that fieldwork is essential to developing a more informed opinion. Having investigated the Dutch experience for a number of years, and after thoroughly reading the vast literature published in English, I went to the Netherlands for one month in the summer of 1999 to get a feel for the local situation. I felt that this would provide the basis on which I could better interpret the findings of the available literature. I visited the major centers of medical ethics, as well as some research hospitals, and spoke with leading figures in the euthanasia policy and practice. The time spent was extremely beneficial and enriching. I followed in the footsteps of Carlos Gomez, who published a book following one month of extensive research in the Netherlands.[1] Two years later, during the summer of 2001, I returned to the Netherlands for a further two weeks to update my research, and in April 2002 I arrived for a further eight days. The present study reports the main findings of my interviews and provides detailed accounts of the way in which some of the Netherlands' leading experts perceive the policy and practice of euthanasia in their country. Their accounts are fascinating. Indeed, the experience was so overwhelming that after I left the Netherlands, I felt that I needed some months to digest the vast information in my mind before sitting down to write the book.

I commence the discussion by addressing the way different democracies view euthanasia and physician-assisted suicide. Attention is given to the familiar distinction between active and passive euthanasia, and then the current legal positions in the Netherlands, Australia, the United States, Switzerland, Belgium, England, France and Canada are analyzed.

The discussion in *Part A* begins with a review of the three major Dutch reports on euthanasia and the conflicting views and interpretations offered by the literature. Two of the reports, published in 1990 and 1995, appeared before my fieldwork to the Netherlands and the interviewees reflected on them. The third report conducted in 2001 and published in 2003 became public after my interviews. After analyzing the three reports I provide some data about the Dutch practice of euthanasia, the legal framework, and the leading court cases.

Part B of the study gives an account of the interviews conducted during the summer of 1999, denoting the similarities and differences between the points of view (*Phase I*).[2] It then reports (in *Phase II*) some of the major comments of the interviewees following their reading of the first draft of this study. I thought it is absolutely essential to let the interviewees have an opportunity to read the manuscript and voice their agreement, and disagreement, about the way their views were presented. I also provided the interviewees with the opportunity to read

1 Carlos F. Gomez, *Regulating Death* (New York: The Free Press, 1991).

2 One interview with Bert Keizer was conducted in April 2002.

the second draft, if they so desired. In turn, *Phase III* gives account of the communications I had with the interviewees during the summer of 2001 and in April 2002, before and during my return to the Netherlands in order to update my research on recent developments, most notably on the new euthanasia law that was enacted in April 2001 and came into force in April 2002 as well as on the work of the regional committees that monitor the practice of euthanasia and physician-assisted suicide.

Most of the discussion in *Phase I* gives voice to the interviewees. I want the readers to judge for themselves about how influential Dutch authorities in the field perceive the policy and practice of euthanasia in their country. My own voice is purposefully limited because I wanted to give the floor mainly to the Dutch, allowing them to describe what is going on in their country. That doesn't mean that I am merely reporting. My opinion is clearly heard when certain ideas emerge that deserve close analysis and criticism. The conclusions are also clear and carefully crafted, aiming to better the situation in the Netherlands for the sake of all concerned.

The analysis indicates that the Dutch Guidelines on the policy and practice of euthanasia are far from being stringent. Indeed they are loose enough to allow non-voluntary (when patients are incompetent) and involuntary euthanasia (when patients are competent and made no request to die). They do not provide effective safeguards against abuse and, in short, they simply do not work. Virtually every Guideline has been breached or violated. This finding reiterates that of Hendin.[3] It should alarm all those who wish to make euthanasia legal. This finding should also press those who, like me, believe that physician-assisted suicide should be made legal, to devise better control mechanisms and establish more workable and effective guidelines. I believe that patients at the end stage of their lives should have the right to decide about the timing of their death.[4] However, there is always difficulty in moving from the recognition of an individual right to the implementation of a policy. The Dutch decision makers have the best of motives, but the policy suffers from serious flaws. The picture that arises from the literature, as well as from the interviews, is disturbing. The Dutch should continue to conduct further investigation and research to improve their policy and practice of euthanasia. Sincere debates and open discussions in which all people who are interested in the subject, from different walks of life, cultures and nationalities, should be encouraged. Exchange of ideas is beneficial to all people concerned. As John Stuart Mill observed, any intolerance of opinions involves, *ipso facto,* a claim to infallible knowledge. Even those opinions of whose truthfulness we are confident, such as "Newtonian philosophy," must be exposed to scrutiny and doubts. Those who assume that they know what the truth is provide reasons against pursuing constant inquiry and debate, depriving humanity of exploring further truths, and so blocking the wheels of progress. Unity of opinion, Mill argued, unless resulting from the fullest and freest comparison of opposite opinions, is not desirable, and diversity not an evil, but a good.[5] According to this view, the quest for truth is a vital as well as an

3 Herbert Hendin, *Seduced by Death* (New York: W.W. Norton, 1997), p. 23.

4 See R. Cohen-Almagor, *The Right to Die with Dignity: An Argument in Ethics, Medicine, and Law* (Piscataway, NJ: Rutgers University Press, 2001). See also the compassionate article of Timothy E. Quill, "Death and Dignity: A Case of Individualized Decision Making," *New. Eng. J. Med.,* Vol. 324, No. 10 (March 7, 1991): 691–694.

5 John Stuart Mill, *Utilitarianism, On Liberty and Representative Government* (London: J. M. Dent, Everyman's edition, 1948), esp. p. 83.

expedient endeavor. Every opinion should be allowed to be aired and then checked against experience, this out of the respect that we feel for the other's freedom of thought and expression; because of the desire to advance the search for the truth; and because debates on different views help individuals become aware of the interests of others, and thus contribute to a sense of community.

I conclude by recommending that the Netherlands amend its policy and remedy its troubling problems by adhering to more rigid Guidelines to prevent potential abuse. Many Dutch scholars take pains to emphasize that there is no slippery slope, and that the euthanasia practice is not deteriorating. Even if this observation is correct, the existing situation is bad enough. Democracies that struggle with the issue, whether to legalize euthanasia or physician-assisted suicide, should learn from the Dutch experience to avoid the pitfalls that are evident in the Netherlands.

PRELIMINARIES: COMPARATIVE LAW

INTRODUCTION

This chapter addresses the way different democracies view the issue of the right to die with dignity. Attention is given to the familiar distinction between active and passive euthanasia, and then the current legal positions in the Netherlands, Australia, the United States, Switzerland, Belgium, England, France and Canada are analyzed. The analysis shows that the movement towards legalizing euthanasia and physician-assisted suicide (PAS) is growing and that there is more legal involvement in end-of-life medical decisions.

Some patients may feel that they are about to die or wish to draw their death nearer while maintaining their dignity. Faced with the deterioration of the functions of their bodies, patients may find it hard to maintain their dignity. Some of them feel exhausted and no longer wish to continue their struggle, especially when they are required (so they feel) to use their energies not only to fight the decaying forces but also against relatives and nursing personnel who sometimes tend to treat them as either infantile or senile, subjects worthy or unworthy of their mercy. The patients' motivations and inclinations help us recognize the right to die with dignity. A distinction should be drawn between the *right* to die with dignity and the *process* of dignified dying.

ACTIVE AND PASSIVE EUTHANASIA

Liberals consider first and foremost the rights and interests of the individual. It has been argued that respecting human life permits, and in some cases argues for, mercy killing (merciful treatment that results in death). In this context, a distinction has been made between *active* and *passive* euthanasia. Euthanasia is a Greek term meaning easy death (*eu* means good, easy; *thanatos* means death). *Active euthanasia* involves prescribing medication or treatment aimed at shortening life and suffering. The attending physician may do it using, for example, a poisonous injection or prescribing large doses of drugs with the intention of cutting short the patient's life. Some describe this action as "killing."[1] *Passive euthanasia* (also termed "forgoing life-sustaining treatment") may take two forms: one is abstention from performing acts that prolong the patient's life. An example may be refraining from connecting a patient to a respirator or to a resuscitation machine. The other form involves discontinuation of actions designed to sustain life. This means withdrawing machines to which the patient has already been connected.[2]

It might be argued that if PAS is sometimes morally permissible, then active euthanasia would also sometimes be morally permissible because both are morally equivalent actions.

[1] For further discussion, see Natalie Abrams, "Active and Passive Euthanasia," *Philosophy*, Vol. 53 (1978): 257–263; Philip Montague, "The Morality of Active and Passive Euthanasia," *Ethics in Science and Medicine*, Vol. 5 (1978): 39–45; Bruce Jennings, "Active Euthanasia and Forgoing Life-Sustaining Treatment: Can We Hold the Line?," *Journal of Pain and Symptom Management*, Vol. 6, No. 5 (July 1991): 312–316.

[2] For further discussion, see James Rachels, "Killing and Letting People Die of Starvation," *Philosophy*, Vol. 54, No. 208 (April 1979): 159–171.

This is the prevalent view in the Netherlands. Undoubtedly the role that the consenting doctors would be expected to play is great and onerous. Doctors who agree with this rationale would argue that while it is true that the doctor's job is to prolong life, his or her job is also to prevent suffering and ensure the preservation of human dignity. Sometimes prolonging life and the preservation of human dignity are mutually exclusive. For a small minority of patients, the continuation of living at all costs is not an appealing option. These patients should not be ignored. Medicine and ethics should address their needs. Although not an easy task, the solution must not be beyond our reach either medically or ethically. That solution might change the nature of medicine but the "nature of medicine" is not a static concept. It is in constant flux and through the ages it has developed by using different standards and norms (see, for instance, the agenda and terminology of the Hippocratic Oath). The history of the last thirty years shows that medicine has changed dramatically due to rapid technological developments. These same developments make it possible to prolong life in difficult situations. An acrobatic argument that acknowledges technological advances but dismisses the evolving ethical issues that pose uncomfortable and disturbing questions is unfair to the community of patients.

COMPARATIVE LAW

Active euthanasia is considered a criminal offense in most countries of the world. The Netherlands exhibits the most permissive attitude in this sphere. Euthanasia and assisted suicide have been practiced and tolerated in the Netherlands over the past twenty years, even though it remained until the passing of the euthanasia law an illegal act under Articles 293 and 294 of the Penal Code. Several lower court decisions, supported by a Supreme Court decision and reflected in the policies of the regional attorneys general and further promulgated by the Royal Dutch Medical Association, have held that when euthanasia meets a certain set of Guidelines, it may be defended under a plea of *force majeure* and so is reasonably sure of being subjected to prosecution.[3] The euthanasia law, which was approved by the Dutch Upper House of parliament on April 10, 2001 places euthanasia and PAS outside the Dutch Penal Code when doctors follow a specified administrative procedure. Euthanasia and PAS would be supervised, not as in the past by the public prosecutor, but by a public committee consisting of a doctor, a lawyer, and an ethics expert. Doctors must be "convinced" that the patient's request is voluntary and well-considered and that the patient is facing "unremitting and unbearable" suffering; have advised the patient of his or her situation and prospects; and have reached a firm conclusion with the patient that there is "no reasonable alternative." At least one other independent physician must have examined the patient. It also ensures that parental consent will now be required before incurably sick minors aged 12 to 16 can request euthanasia.[4] I will elaborate on the legal framework in *Chapter 2*.

3 In February 1993 the Second Chamber of the Dutch Parliament accepted these provisions. See Margaret P. Battin, *The Least Worst Death* (New York: Oxford University Press, 1994), pp. 130–131. See also Sjef Gevers, "Physician Assisted Suicide: New Developments in the Netherlands," *Bioethics*, Vol. 9, Nos. 3/4 (1995): 309–312, and RDMA, *Euthanasia in the Netherlands* (4th Edition) (Utrecht, December 1995).

4 See "The Netherlands: Bill on Euthanasia and Assisting Suicide in the Netherlands," *European J. of Health Law*, Vol. 5 (1998): 299–324; Tony Sheldon, "Netherlands Gives More Protection to Doctors in Euthanasia

In Australia, the Legislative Assembly of the Northern Territory passed on May 25, 1995 the Northern Territory *Rights of the Terminally Ill Act 1995,* allowing terminally ill patients to commit suicide with a doctor's help. The legislation, applied only in the Northern Territory, enabled a terminally ill Australian adult, experiencing "unacceptable" pain, to be examined by a qualified physician to determine whether the patient could be cured.[5] The Act required confirming examinations by two other independent physicians, one specializing in treating terminal illness and the other a qualified psychiatrist, to confirm that the patient is terminally ill and not clinically depressed.[6] After considering the advice of the consultants, medical assistance to die could not take place if there were palliative care options "reasonably available to the patient to alleviate the patient's pain and suffering to levels acceptable to the patient."[7]

The Act also included provisions intended to ensure that the patient was making an informed choice. The doctor had to have informed the patient of the nature of the illness, its likely course and the medical procedures available. Upon having the pertinent information, the patient could indicate that he or she wishes to end his or her life. The doctor had to be satisfied that the patient had considered the possible implications of the decision for his or her family, and that the decision had been made freely, voluntarily and after due consideration.[8]

The Act provided for a nine-day "cooling off" period comprised of two stages. Seven days had to elapse between the initial request and the signing of the certificate of request and a further 48 hours before providing assistance to terminate life.[9] The signing of the certificate had to be witnessed by two doctors. If the patient was physically unable to sign the certificate, a person other than the doctors and psychiatrist referred to, who was at least 18 years old, could sign it on behalf of the patient. The person could not be likely to receive any financial benefit from the patient's death, and forfeited any benefit if he or she would in fact have received it.[10] The statute provided that the patient could rescind at any time and in any manner,[11] and the physician was under no obligation to assist suicide. If the physician chose to comply with the patient's request, death could be hastened by prescribing or preparing a lethal substance, giving the substance to the patient for self-administration, or administering the lethal substance to the patient. The doctor had to remain present until the death of the patient.[12]

The legislation became operative in July 1996, and in the following nine months four patients who requested to die received help under the provisions of the Act by Dr. Philip

Cases," *British Medical Journal*, Vol. 321 (December 9, 2000), 1433; Rory Watson, "MEPs Try to Mobilise Public Opinion against Extension of Euthanasia," *British Medical Journal*, Vol. 322 (March 17, 2001): 638.

5 *Rights of the Terminally Ill Act 1995* (NT), Section 4.

6 *Ibid*, Section 7.

7 *Ibid*, Section 8.

8 *Ibid*, Section 7.

9 *Ibid*, Section 7.

10 *Ibid*, Section 9.

11 *Ibid*, Section 10.

12 *Ibid*, Section 7. For further discussion, see Simon Chesterman, "Last Rights: Euthanasia, the Sanctity of Life, and the Law in the Netherlands and the Northern Territory of Australia," *International and Comparative L. Q.*, Vol. 47 (April 1998), esp. pp. 386–387; Andrew L. Plattner, "Australia's Northern Territory: The First Jurisdiction to Legislate Voluntary Euthanasia, and the First to Repeal It," *DePaul J. of Health Care Law*, Vol. 1 (Spring 1997), esp. pp. 647–648.

Nitschke.[13] The Act was annulled in March 1997, when federal parliamentarians by 38 votes to 34, with one abstention, passed the *Commonwealth Euthanasia Laws Bill 1996.*[14] That Act effectively prohibits Australian territories from enacting legislation that permits "the form of intentional killing of another called euthanasia… or the assisting of a person to terminate his or her life," but allows the making of laws regarding the withdrawal or withholding of life-sustaining treatment and the provision of palliative care to the dying, provided these do not sanction the intentional killing of the patient.[15] After Federal Parliament overturned the Northern Territory's euthanasia law, Dr. Philip Nitschke revealed that he had helped 15 patients, including some from Victoria, to end their lives.[16]

In the United States, attempts made in 1988–1992 in the states of Washington[17] and California[18] to pass laws recognizing the possibility of active euthanasia were unsuccessful. In a 1998 referendum, Michigan voters overwhelmingly (70% to 30%) rejected the legalization of physician-assisted suicide, and in November 2000, Maine voters had done the same, failing the ballot measure to legalize PAS by a narrow 51 percent to 49 percent decision.[19] In May 2002, Hawaii State Senate defeated a bill to allow assisted suicide when three lawmakers switched their votes. The Senate voted 14–11 to reject House Bill 2487, which would have allowed terminally ill, competent people to obtain lethal prescriptions that they could take themselves to end their lives.[20]

In November 1994, voters in Oregon approved the first American law allowing doctors to hasten death for the terminally ill. The Oregon *Death with Dignity Act* (Measure No. 16) was designed to protect the following interests of the patients: avoiding unnecessary pain and suffering; preserving and enhancing the right of competent adults to make their own critical health care decisions; avoiding tragic cases of attempted or successful suicides in a less humane and dignified manner; protecting "the terminally ill and their loved ones" from financial hardships they wish to avoid, and protecting "the terminally ill and their loved ones

[13] Philip Nitschke, "Do No Harm," *Family Circle* (April 1, 1998), p. 126; Gay Alcorn, "First death under NT mercy law," *The Age* (Melbourne) (Sept. 27, 1996). On Dr. Nitschke and his campaign for euthanasia in Darwin and Australia, see "Australia has its own 'Kevorkian'," *The Associated Press* (January 11, 1998); film *The Road to Nowhere*, Four Corners, ABC (broadcast in Australia on July 8, 1996). See also URL of the South Australian Voluntary Euthanasia Society (SAVES): http://www.on.net/clients/saves

[14] See the film *Where Angels Fear to Tread,* produced by the Australian Film Finance Corporation Ltd. and Annamax Media Pty. Ltd. (broadcast on the Science Channel No. 8 in Israel on November 4, 1998).

[15] Helga Kuhse, "From Intention to Consent," in Margaret P. Battin, Rosamond Rhodes and Anita Silvers (eds.), *Physician Assisted Suicide* (New York and London: Routledge, 1998), p. 252.

[16] Darren Gray, "Doctor: I helped 15 patients die," *The Age* (Melbourne) (November 27, 1998). Chris Ryan, "Right-to-die bill pleases doctor," *The Age* (Melbourne) (July 11, 1997). Further information is available from: Hon. Secretary, SAVES, PO Box 2151, Kent Town, SA 5071, Australia – Fax + 61 8 8265 2287. For further deliberation on end-of-life practice in Australia, see H. Kuhse, P. Singer, P. Baume *et al.*, "End-of-Life Decisions in Australian Medical Practice," *Medical J. of Australia* (*MJA*), Vol. 166 (1997): 191–196; Colleen M. Cartwright, "Attitudes of Australian Doctors, Nurses and Community Members towards Physician-Assisted Suicide and Euthanasia," in Agnes van der Heide *et al.* (eds.), *Clinical and Epidemiological Aspects of End-of-Life Decision-Making* (Amsterdam: Proceedings of colloquium, 2001): 135–147; Roger S. Magnusson, *Angels of Death* (Carlton South, Victoria: Melbourne University Press, 2002).

[17] Initiative for Death with Dignity, Washington Initiative No. 119 (1991).

[18] The California Death with Dignity Act, California Proposition No. 161 (1992).

[19] Tim Christie, "Voters in Maine Reject Assisted-Suicide Law," *The Register-Guard* (Eugene, Ore.) (November 14, 2000). See also E.J. Emanuel, "Euthanasia and Physician-assisted Suicide – A Review of the Empirical Data from the United States," *Archives of Internal Medicine*, Vol. 162 (January 28, 2002): 142–152.

[20] Lynda Arakawa and Kevin Dayton, "Assisted suicide bill rejected," *The Honolulu Advertiser* (May 3, 2002).

from unwanted intrusions into their personal affairs" by law enforcement officers and others.[21]

In 1994, the Oregon law was approved by a 51 percent to 49 percent vote of the state residents,[22] but was promptly put on hold amid great legal wrangling. Two days before the *Death with Dignity Act* was to take effect, a lawsuit was filed by a group of physicians, residential care facilities, and terminally ill Oregon residents challenging the Act on constitutional grounds.[23] The federal district court granted a temporary injunction and eight months later struck down the Act on equal protection grounds.[24]

This decision was subsequently vacated for procedural reasons, remanding the judgment of the district court for lack of jurisdiction.[25] Then, in 1997, the state's voters backed the law again, this time by a decisive margin of 60 percent to 40 percent.

The Oregon *Death With Dignity Act* allows Oregonian patients who suffer from a terminal disease and who have a life expectancy of no more than six months to receive prescriptions for self-administered lethal medications from their physicians. The Act legalizes only physician-assisted suicide, stating that "nothing in this Act shall be construed to authorize a physician or any other person to end a patient's life by lethal injection, mercy killing or active euthanasia."[26] It permits a capable[27] adult[28] Oregon resident[29] diagnosed with a terminal disease to make a written request for medication for the purpose of ending his or her life in a humane and dignified manner. Physicians who write such prescriptions in good-faith compliance with the Act are shielded from civil or criminal penalties and professional discipline.[30]

In response to concerns about inadequate safeguards, the authors of the Oregon *Death With Dignity Act* provided detailed procedures that patients and physicians must follow. The patient who is suffering from a terminal disease must first make an oral request, then a written request[31] and lastly an additional oral request before the "attending physician"[32] may assist. The written request must be signed and dated by the patient, and witnessed by at least two individuals one of whom must not be a relative, an heir, or the owner or operator of a

21 Defendants' Reply Memorandum in Support of Motion for Summary Judgment, *Garry Lee v. State of Oregon*, Civil No. 94-6467-HO, United States District Court (August 3, 1995), at 9.

22 The voter turnout was 57 percent. A breakdown of the results showed that the vote was 55 percent for and 45 percent against in the affluent areas. It lost in most rural counties that have smaller populations. Mark O'Keefe, "Assisted-suicide measure survives," *The Oregonian* (November 10, 1994), at A1.

23 *Lee v. Oregon*, 869 F.Supp. 1491 (D. Oregon 1994).

24 *Lee v. Oregon*, 891 F.Supp. 1429 (D. Oregon 1995). For criticism of this court ruling, see Charles H. Baron, Clyde Bergstresser, Dan W. Brock *et al.*, "A Model State Act to Authorize and Regulate Physician-Assisted Suicide," *Harvard J. of Legislation*, Vol. 33, No. 1 (1996): 14–16.

25 *Lee v. Oregon*, 107 F.3d 1382, 1392 (9th Cir. 1997).

26 The Oregon *Death With Dignity Act* § 3.14, *Or. Rev. Stat.* § 127.880 (1999).

27 "Capable" is defined as having the ability to make and communicate health care decisions to a health care provider. See Oregon *Death With Dignity Act* § 1.01(3).

28 An "adult" is an individual who is at least eighteen years of age.

29 Oregon *Death With Dignity Act* § 3.10. The residency requirement was intended to prevent individuals from other states from rushing to Oregon to take advantage of the Act.

30\. Oregon *Death With Dignity Act* § 4.01(1). Without this provision, physicians who assisted a suicide would be civilly and criminally liable under *Or. Rev. Stat.* § 163.125 (1995), which makes it a crime to intentionally cause or aid another in committing suicide.

31 The statute contains a form for the written request, and requires that two witnesses affirm that the patient is capable and is acting voluntarily in making the request. See Oregon *Death With Dignity Act* §§ 2.02, 6.01.

32 The "attending physician" is the doctor with primary responsibility for the care of the patient.

health care facility where the patient is receiving treatment or is in residence. Neither of the witnesses shall be the patient's attending physician. The requirement of both oral and written requests encourages the patient to consider his or her condition and the significance of the decision, thus serving the best interests of the patient. It also provides physicians with a record of the patient's wishes, to safeguard them from liability. In order to ensure that the patient's request is not a result of familial pressure, I suggest that the doctor or another member of the medical team will be obliged to conduct conversations with the patient and the relative to see that their motives are genuine, aiming to serve the patient's best interests. The witness should not be a beneficiary with partisan interests that might cloud his/her judgment.[33]

Physicians must also allow the patient to withdraw his or her request at any point, and are required to explicitly offer the patient the opportunity to change his or her mind before prescribing a lethal dose of medication.[34] Like the requirement for both oral and written requests, these requirements provide additional safeguards to ensure that the patient is making a voluntary, informed, and cautious decision. They likewise protect the best interests of patients by encouraging them to reconsider their choice and provide prescribing physicians another indication that the patient is not making a rash or coerced decision.

Once the patient makes the first oral request, the physician must inform the patient of the diagnosis, prognosis, potential risks and probable result of taking the prescription, as well as alternatives including pain management and comfort and hospice care.[35] This ensures that the patient could make a reasoned and informed decision.

The physician must wait at least fifteen days after the patient's first oral request before writing the prescription,[36] arguably too long for a patient who is on the verge of death. Although it is important to allow sufficient time for the patient to contemplate his/her decision and for physicians to assess the patient, establishing a fixed waiting period may prevent a patient closest to death from utilizing PAS before his or her natural death. During the waiting period, the attending physician must consult another physician[37] and ask the patient to notify next of kin regarding the decision. The referral to a consulting physician prevents one physician from making a unilateral decision to prescribe lethal medication. It also allows an important additional evaluation of the patient's illness, prognosis, and mental soundness.

The Oregon Act is a significant step toward establishing a patient's right to autonomy and choice in deciding end of life issues. The findings of the Oregon Health Division portray the people opting for assisted suicide as well-educated, well-insured, often in hospice care, and very concerned about loss of independence. Worries about money played essentially no role in the patients' decision. There is no evidence that the poor, uneducated, mentally ill, or

33 See R. Cohen-Almagor and Monica G. Hartman, "The Oregon Death with Dignity Act: Review and Proposals for Improvement," *Journal of Legislation,* Vol. 27, No. 2 (2001): 269–298.

34 Oregon *Death With Dignity Act* § 3.07.

35 Oregon *Death With Dignity Act* § 3.01(2)(a)-(e).

36 Oregon *Death With Dignity Act* § 3.08.

37 The "consulting physician" is the doctor qualified by specialty or experience to render a professional diagnosis and prognosis about the patient's condition.

socially isolated are disproportionately seeking or getting lethal prescriptions of drugs under the Oregon *Death with Dignity Act.*[38]

Following the Oregon *Death with Dignity Act*, attempts to legalize assisted suicide were made by many other states, among them Alaska,[39] Arizona, California, Colorado, Connecticut, Illinois, Massachusetts, Michigan, Mississippi, Nebraska, Washington and Wisconsin. All attempts were rejected.[40] A bigger setback for assisted suicide advocates came in the summer of 1997, when the US Supreme Court upheld laws in Washington State and New York banning physician-assisted suicides. The Court reversed decisions of the lower courts that held those laws unconstitutional.

In 1994, the United States District Court was asked to rule on the constitutionality of the State of Washington's criminal prohibition against physician-assisted suicide. Specifically, the plaintiffs asserted that the Fourteenth Amendment to the Constitution guarantees mentally competent, terminally ill adults who act under no undue influence the right to voluntarily hasten their death by taking a lethal dose of physician-prescribed drugs. Chief Judge Rothstein held that such a right for these patients was liberty interest protected under the Fourteenth Amendment, and that the Washington 140-year-old anti-assisted-suicide statute violated equal protection by prohibiting these patients from seeking physician-assisted suicide but allowing withdrawal of life-support systems.[41] Later, this judgment was reversed by the United States Court of Appeals, Ninth Circuit, *per* Judge Noonan, saying that the Washington statute did not deprive patients of constitutionally protected liberty interest.

[38] Amy D. Sullivan, Katrina Hedberg and David W. Fleming, "Legalized Physician-assisted Suicide in Oregon – the Second Year," *New Eng. J. of Med.*, Vol. 342, No. 8 (February 24, 2000): 598–604. See also David Brown, "A picture of assisted suicide: most who use Oregon law are educated, insured; some change their minds," *Washington Post* (February 24, 2000), p. A03.

[39] In September 2001, the Alaska Supreme Court ruled that Alaskans have no constitutional right to assisted suicide, despite a constitution and courts that strongly uphold individual liberty and personal privacy. In a unanimous decision the court upheld a state law that says helping others kill themselves is manslaughter. The court cited the state's long history of opposition to assisted suicide and the danger that vulnerable Alaskans who are depressed or disabled might be pressured into dying when they do not really want to. Cf. Sheila Toomey and Ann Potempa, "Court Upholds State Law Barring Assisted Suicide," *Anchorage Daily News* (September 22, 2001). http://www.adn.com/

[40] Assisted suicide is a crime by statute in the following states: Alaska, Arizona, Arkansas, California, Colorado, Connecticut, Delaware, Florida, Georgia, Hawaii, Illinois, Indiana, Iowa, Kansas, Kentucky, Louisiana, Maine, Minnesota, Mississippi, Missouri, Montana, Nebraska, New Hampshire, New Jersey, New Mexico, New York, North Dakota, Oklahoma, Pennsylvania, Rhode Island, South Dakota, Tennessee, Texas, Washington and Wisconsin. Common law forbids assisted suicide in: Alabama, Idaho, Massachusetts, Nevada, Vermont and West Virginia. States in which physician-assisted suicide is considered a criminal act through statutes and common law are: Maryland, Michigan and South Carolina. States without common laws or statute laws on assisted suicide are North Carolina, Utah, and Wyoming. In addition, Virginia has no clear case law nor any statute on assisted suicide but does have a state statute that imposes civil sanctions on persons assisting in a suicide. And the Ohio Supreme Court ruled in October 1996 that assisted suicide is not a crime. Source: National Conference of State Legislatures (November 1997), reported by The Associated Press, Status of Assisted Suicide by State (January 5, 1998), and by American Medical News (May 2000). URL: http://www.ama-assn.org/amednews/2000/05/22/prsb0522.htm. For discussion on the legislative attempts to legalize PAS in ten states during 1997, see Russell Korobkin, "Physician-assisted Suicide Legislation: Issues and Preliminary Responses," *Notre Dame J. of Law, Ethics & Public Policy*, Vol. 12, No. 2 (1998): 449–472.

[41] *Compassion in Dying v. State of Washington*, 850 F. Supp. 1454 (W. D. Wash. 1994). For further discussion, see Ronald Dworkin, *Freedom's Law* (Cambridge, Mass. Harvard University Press, 1996):143–146.

The panel majority found that the statute prohibiting suicide promotion furthered, *inter alia*, the interest in denying to physicians "the role of killers of their patients."[42]

However, physicians do not fulfill the role of "killers" by prescribing drugs to hasten the death of patients who voluntarily chose this option more than they do by withdrawing life-support machines. The court clouded an important issue by resorting to this radical language instead of probing the complexities involved and establishing procedures to ensure patients against possible abuse.

The Ninth Circuit Court reheard the case, reversed the panel decision, and affirmed the District Court ruling.[43] Circuit Judge Reinhardt opened his judgment with the following thoughtful words, which are quoted in full:

> This case raises an extraordinarily important and difficult issue. It compels us to address questions to which there are no easy or simple answers, at law or otherwise. It requires us to confront the most basic of human concerns -- the mortality of self and loved ones -- and to balance the interest in preserving human life against the desire to die peacefully and with dignity. People of good will can and do passionately disagree about the proper result, perhaps even more intensely than they part ways over the constitutionality of restricting a woman's right to have an abortion. Heated though the debate may be, we must determine whether and how the United States Constitution applies to the controversy before us, a controversy that may touch more people more profoundly than any other issue the courts will face in the foreseeable future.[44]

The court held that there was a constitutionally protected liberty interest in determining the time and manner of one's own death, an interest that must be weighed against the state's legitimate and countervailing interests, especially those related to the preservation of human life. After balancing the competing interests, the court concluded that insofar as the Washington statute prohibited physicians from prescribing life-ending medication for use by terminally ill, competent adults who wish to hasten their deaths, it violated the Due Process Clause of the Fourteenth Amendment.[45]

Finally, the Supreme Court, *per* Chief Justice Rehnquist, reversed the judgment yet again, holding that the asserted right to assistance in committing suicide was not fundamental liberty interest protected by the Due Process Clause, and that Washington's ban on assisted suicide was rationally related to legitimate government interests, among them the preservation of human life; the protection of the integrity and ethics of the medical profession; the protection of vulnerable groups from abuse, neglect, and mistakes; and furthermore, the protection of disabled and terminally ill people from prejudice, negative and inaccurate stereotypes, and societal indifference.[46]

Interestingly, Chief Justice Rehnquist chose to conclude his opinion by calling on the public to continue the debate in earnest:

> Throughout the Nation, Americans are engaged in an earnest and profound debate about the morality, legality, and practicality of physician-assisted suicide. Our holding permits this debate to continue, as it should in a democratic society.[47]

42 *Compassion in Dying v. State of Washington*, 49 F. 3d 586 (9th Cir. 1995), at 592.
43 *Compassion in Dying v. State of Washington*, 79 F. 3d 790 (9th Cir. 1996).
44 *Compassion in Dying v. State of Washington*, 96 C.D.O.S. 1507, at 2.
45 *Ibid.*
46 *Washington v. Glucksberg*, 117 S. Ct. 2258 (1997).
47 *Ibid*, at 2275.

In 1996, Dr. Timothy Quill and his colleagues appealed to the courts to declare two New York statutes penalizing assistance in suicide unconstitutional in part. The physicians argued that the statutes under examination were invalid to the extent that they prohibited them from acceding to the requests of terminally ill, mentally competent patients for help in hastening death. The court struck down the statutes, finding that they violated the Equal Protection Clause of the Fourteenth Amendment because they were not "rationally related to a legitimate state interest." Quoting a series of precedents from 1914 onwards, the court said that the right to refuse medical treatment has long been recognized in New York,[48] holding that physicians who are willing to do so may prescribe drugs to be self-administered by mentally competent patients who seek to end their lives during the final stages of a terminal illness.[49]

Representatives of New York appealed to the Supreme Court that reversed the decision. The Court, *per* Chief Justice Rehnquist, held that New York's prohibition on assisted suicide did not violate the Equal Protection Clause of the Fourteenth Amendment. The Court maintained that the distinction between letting a patient die and making that patient die is important, logical, rational and well established.[50]

As a result of the two decisions, *Washington v. Glucksberg* and *Vacco v. Quill,* both given on June 26, 1997, by a 9 to 0 vote, the states have responsibility for insuring that the interests of all patients near the end of their lives are not imperilled. In the American pluralistic society there are widely differing values and sharply clashing views on mercy killings, physician-assisted suicide and the right to die with dignity. I mentioned Measure 16 enacted in Oregon. In Michigan, a special statute was passed in 1993 to stop Dr. Jack Kevorkian from assisting patients to die.[51] However, Dr. Kevorkian stood trial several times and in all instances the juries refused to convict him of violating that statute, although Kevorkian admitted he had assisted people to commit suicide. Kevorkian was acquitted on the grounds that his main intent was to relieve pain, not to cause death.[52] In November 1998, he actively performed euthanasia on Thomas Youk, stood trial and was convicted of second-degree murder and for delivering a controlled substance for the purpose of injecting Youk

48 *Schloendorff v. Society of New York Hosp.*, 211 N.Y. 125, 129 (1914); *In re Storar*, 52 NY2d 363, 420 NE2d 64, cert denied, 454 U.S. 858, 70 L.Ed 2d 153, 102 S.Ct. 309 (1981); *In re Eichner* 52 N.Y. 2d 363, 438 N.Y.S. 2d 266, 420 N.E. 2d 64 (1981); *Rivers v. Katz*, 67 N.Y. 2d 485 (1986).

49 *Quill v. Vacco*, U.S. Court of Appeals for the Second Circuit (April 2, 1996), No. 95-7028, at 12.

50 *Vacco v. Quill* 117 S.Ct. 2293 (1997), at 2295, 2301. For analysis of *Washington v. Glucksberg, Compassion in Dying v. State of Washington*, and *Vacco v. Quill*, see Yale Kamisar, "On the Meaning and Impact of the Physician-Assisted Suicide Cases," *Minnesota L. Rev.*, Vol. 82 (1998): 895–922; Cass R. Sunstein, *One Case at a Time* (Cambridge, Massachusetts: Harvard University Press, 1999): 75–116. For an overview of the legalization efforts of physician-assisted suicide in the US, see Carol A. Pratt, "Efforts to Legalize Physician-Assisted Suicide in New York, Washington and Oregon: A Contrast between Judicial and Initiative Approaches – Who Should Decide?," *Oregon L. Rev.*, Vol. 77 (Winter 1998): 1027–1123.

51 Mich. Comp. Laws Ann. 752.1027 (West Supp. 1995). See Janet M. Branigan, "Michigan Struggle with Assisted Suicide and Related Issues as Illuminated by Current Case law: An Overview of *People v. Kevorkian*," *Un. Of Detroit Mercy L. Rev.*, Vol. 72 (1995), at 959–960.

52 *State of Michigan v. Kevorkian*, Michigan CirCt (Oakland City), verdict March 8, 1996. Kevorkian outlines his rationale for assisted suicide in *Prescription: Medicide* (New York: Prometheus Books, 1991). For further deliberation on developments in American law, see David Orentlicher, "The Legalization of Physician Assisted Suicide: A Very Modest Revolution," *Boston College L. Rev.*, Vol. XXXVIII, No. 3 (May 1997): 443–475.

with lethal drugs. Kevorkian was given a jail sentence of 10 to 25 years on the 2nd degree murder conviction, and 3 to 7 years on the "controlled substance" conviction.[53]

In November 2001 Attorney General John Ashcroft declared that doctors may not prescribe lethal doses of federally controlled substances to terminally ill patients. Overturning the policy adopted by Attorney General Janet Reno (Reno wrote an opinion in 1998 saying that the federal drug law was never intended to regulate medical practice), Ashcroft sided with the Drug Enforcement Administration (DEA), which had long argued that doctors who prescribe drugs under Oregon's assisted-suicide law could lose their licenses to write prescriptions.[54] While physicians are licensed by the states to practice medicine, the DEA registers doctors to prescribe drugs and the agency is responsible for enforcing the federal controlled substances law. Ashcroft determined that "assisting suicide is not a 'legitimate medical purpose' within the meaning of (the law) and that prescribing, dispensing or administering federally controlled substances to assist suicide" violates federal law.[55]

The state of Oregon was quick to respond and filed motions in U.S. District Court seeking to temporarily prevent the federal government from implementing the new order barring doctors from prescribing federally controlled substances to hasten the deaths of terminally ill patients. Attorney General Hardy Myers also filed a lawsuit challenging the authority of U.S. Attorney General Ashcroft to limit the practice of medicine in Oregon by attempting to bar physician-assisted suicides.[56]

In April 2002, a federal judge ruled that the U.S. Justice Department lacks the authority to overturn a voter-approved Oregon law allowing physician-assisted suicides. U.S. District Judge Robert Jones scolded Attorney General John Ashcroft, saying Ashcroft with no advance warning to Oregon "fired the first shot in the battle between the state of Oregon and the federal government over which government has the ultimate authority to decide what constitutes the legitimate practice of medicine."[57] Jones said Ashcroft attempted to "stifle an ongoing, 'earnest and profound debate' in the various states concerning physician-assisted suicide" with a directive declaring that assisted suicide was not a "legitimate medical practice."[58] However, nothing in the plain language of the *Controlled Substances Act* (CSA) or its legislative history "demonstrates Congress' intent to grant defendants the authority under the CSA to determine that prescribing controlled substances for purposes of physician-assisted suicide in compliance with Oregon law is not a 'legitimate medical purpose'."[59] He maintained: "As the Court acknowledged in *Glucksberg,* the citizens of Oregon, through

[53] For further deliberation on Kevorkian, see R. Cohen-Almagor, *The Right to Die with Dignity: An Argument in Ethics, Medicine, and Law* (Piscataway, NJ.: Rutgers University Press, 2001), pp. 191–200.

[54] Republican leaders in Congress complained that Reno had acted in secret, and her memo ignited legislative battles that flared throughout the 106th Congress of 1999–2000. However, the Bush administration's internal debate over the Oregon law took place entirely behind closed doors and with the participation of few outsiders. Cf. Jim Barnett, "Bush policy on suicide is shielded in secrecy," *The Sunday Oregonian* (November 11, 2001).

[55] James Vicini, "Federal doctor-assisted suicide policy reversed," *Reuters* (November 6, 2001); Sam Howe Verhovek, "U.S. Acts to stop assisted suicides," *New York Times* (November 7, 2001).

[56] Brad Cain, "Oregon files lawsuit against U.S. government over assisted suicide," *The Associated Press* (November 7, 2001).

[57] *State of Oregon v. Ashcroft*, US District Court for the District of Oregon, Civil No. 01-1647-JO (April 17, 2002), pp. 2–3.

[58] *Ibid*, p. 3.

[59] *Ibid*, pp. 18–19.

their democratic initiative process, have chosen to resolve the moral, legal and ethical debate on physician-assisted suicide for themselves by voting -- not once, but twice -- in favor of the Oregon act."[60] To allow an attorney general – "an appointed executive whose tenure depends entirely on whatever administration occupies the White House – to determine the legitimacy of a particular medical practice without a specific congressional grant of such authority would be unprecedented and extraordinary."[61] The battle continues.

While the USA struggles with controversy regarding the legality of assisted suicide, the Swiss legal system has condoned the practice for more than sixty years. In contrast to practices in the Netherlands, Australia and the various US proposals where assisted suicide is limited to physician-assisted suicide, Swiss law permits aiding the dying by laypersons.[62] Since 1937, Articles 114 and 115 of the Penal Code have governed assistance to suicide.[63] Although under Article 114 anyone taking another person's life, even if for honorable motives of compassion and at the request of the sufferer, is liable to imprisonment,[64] still the judge will consider whether the patient is terminal, whether he requested death, and whether the patient acted for noble reasons.[65] This is under Article 63 of the Penal Code of 1937, which instructs the judge to mete out punishment in accordance with the guilt of the actor, considering the motives, the prior life, and the personal circumstances of the guilty person. Swiss laws stipulate that persons who assist a suicide do so for humane reasons with no chance of personal gain. The Swiss Penal Code says, "assisted suicide is not punishable, provided it is not done for enrichment purposes and that the person [who wishes to commit suicide] carries out the final death act."[66] Swiss law permits both passive euthanasia – deliberate renunciation or interruption of measures to preserve life, and active indirect euthanasia – administering of substances to reduce suffering whose secondary effects may reduce the period of the patient's survival.[67]

60 *Ibid*, p. 4.

61 *Ibid*, p. 27.

62 S.A. Hurst and A. Mauron, "Assisted Suicide and Euthanasia in Switzerland: Allowing a Role for Non-physicians," *British Medical Journal*, Vol. 326 (February 1, 2003): 271–273.

63 Article 115 of the Swiss Penal Code deals with inciting and assisting suicide. It holds: "Anyone with a selfish motive who incites a person to commit suicide or who helps that person to commit suicide, if the suicide is consummated or attempted, will be punished by a maximum of five years reclusion or imprisonment." If there is no selfish motive, assisted suicide is legal. I am grateful to EXIT A.D.M.D Suisse Romande for sending me the information. E mail: exit@freemail.ch.

64 "Experts want Switzerland as first nation with legal euthanasia," *right_to_die@efn.org*, E-mail sent on April 30, 1999.

65 An initiative proposal of a group of experts from the Federal Assisted Death Commission wishes to add a section to Article 114 that would read: "If the perpetrator helps a person, who is in the final stages of an incurable illness, to die in order to end insupportable and incurable suffering, the competent authority will not proceed against this person, will not force him to appear before a court nor inflict a penalty." In Dr. Jérome Soble, "Assisted Death," trans. from "Revue médicale de la Suisse romande," Vol. 121 (2001): 163–164. Note that according to this proposal anybody can provide assisted suicide, not only people of the medical profession.

66 Derek Humphry, *The Right to Die* (New York: Harper and Row, 1986), pp. 221–222; http://www.finalexit.org/

67 "Experts want Switzerland as first nation with legal euthanasia," *right_to_die@efn.org* (April 30, 1999); "Swiss assisted suicide policy draws attention at euthanasia meeting," APF news story (October 15, 1998). For a comparative study, see G. Bosshard, S. Fischer and W. Bar, "Open Regulation and Practice in Assisted Dying – How Switzerland Compares with the Netherlands and Oregon," *Swiss Medical Weekly*, Vol. 132 (October 12, 2002): 527–534.

Although Swiss law permits physicians and non-medical persons to assist suicides, the Swiss Academy of Medical Sciences, like many medical organizations, including the American Medical Association, opposes doctors helping patients to die.[68]

Today there are no less than four "right-to-die" organizations in a country with a population of seven million. The largest organization is EXIT, the Swiss Society for Human Dying, with 50,000 members, based in Zurich. The other three are much smaller in comparison: EXIT International[69]; Dignitas, with about 1,700 members, of which 1,000 are foreigners, and EXIT A.D.M.D. Suisse Romande, with about 10,000 members.[70] Two of them offer help with hastened death to both terminally ill and chronically ill, be they resident or foreigner. One group will even visit a dying patient in another country if there is no alternative. The main organization, EXIT, requires that the applicant wishing to die with EXIT's help be at least eighteen years old, a Swiss resident, mentally competent and suffering from intolerable health problems. He or she must personally apply for the service and convince the administrators of EXIT that there is no coercion or third party influence involved in the decision. An EXIT physician considers the application and decides whether or not assistance can be offered. In doubtful cases, a team composed of a lawyer, psychiatrist and a physician will make the decision jointly. Once a decision is made, the patient determines the date for his or her death. A member of EXIT team of suicide helpers then comes and brings the medication. The patient has to drink it by him/herself. EXIT welcomes the presence of family members or the patient's loved one's. As soon as death is confirmed, the EXIT member calls the police. According to the Swiss law every unusual death has to be examined by the police to make sure that no crime was committed.[71]

Between 100 and 120 people are openly helped in this way each year. EXIT performed 124 assisted suicides in 2001 (96 in 2000) and provided consultation about assisted suicide to about 1,000 people.[72] So far, no member of an EXIT team has had to appear before the court for helping a person to commit suicide. It is estimated that the four "right-to-die"

68 The Swiss Academy of Medical Sciences takes the view that helping a patient to commit suicide is beyond the scope of medical practice. This makes it difficult to use the law, given that the doctor is the person most qualified to give help to a terminally ill person who wants assistance in suicide. South Australian Voluntary Euthanasia Society, DID YOU KNOW? Assisted Suicide in Switzerland – SAVES Fact Sheet No. 20, issued February 1997. Correspondence with: Hon. Secretary, SAVES, PO Box 2151, Kent Town, SA 5071, Australia – Fax + 61 8 8265 2287.

69 Contact EXIT.-International. Postfach 605, CH-3000, Bern 9. Switzerland.

70 See www.exit.ch; http://www.dignitas.ch/; www.exit-geneve.ch; www.dghs.de; www.finalexit.org/pract-swiss.html.

71 "The Practice of Assisted Suicide in Switzerland," A Report by Professor Meinrad Schaer, MD, President of "EXIT," Society for Human Dying, Switzerland. http://www.finalexit.org/swissframe.html. See also Jérome Sobel, "Assisted Suicide," Exit A.D.M.D. I am grateful to Michael Brucker and Sandrine Rohmer for arranging this information.

72 World Right-to-Die *Newsletter*, Issue 42 (November 2002), p. 7. Cf. www.worldrtd.org

organizations hasten some 200 deaths each year.[73] EXIT A.D.M.D. performed some 40 assisted suicides in 2001 (20 in 2000). Dignitas performed around 50 cases in 2001.[74]

Belgium has followed the euthanasia path of its Dutch neighbor. For some time, there were no formal registration and authorization procedure for end-of-life decisions in medical practice. Although euthanasia was illegal and treated as intentionally causing death under criminal law, prosecutions were exceptional and – generally speaking – the practice of euthanasia was tolerated. Since 1996, legalization of euthanasia was intensely discussed, both by the official Advisory Committee on Bioethics and the Belgian Parliament.[75] Proposals to remove euthanasia from the criminal law had angered doctors, who claimed they had not been properly consulted. Dr. Marc Moens, chairperson of the Belgian Association of Doctors Syndicates (BVAS), which comprises two thirds of the country's 40,000 doctors, argued that abolishing the law on euthanasia would do nothing to prevent abuses but would make "The exception the rule."[76] A study conducted in Flanders showed that the frequency of deaths preceded by an end-of-life decision is similar to that in the Netherlands, but lower than that in Australia. However, in Flanders the rate of administration of lethal drugs to patients without their explicit request is similar to Australia, and significantly higher than that in the Netherlands.[77]

On January 20, 2001 the euthanasia commission of Belgium's upper house, the Senate, voted in favor of proposed euthanasia legislation, which would make euthanasia no longer punishable by law, provided certain requirements are met.[78] Nine months later, on October 25, 2001 Belgium's Senate approved the law proposal, which was adopted on March 20, 2001 by the joint commissions of Justice and Social Affairs, by a significant majority: 44 for, 23 against, 2 abstentions and two senators who failed to register a vote. It was clear beforehand that there was general support among all six parties in the ruling coalition of Socialists, Liberals and Ecologists. In society at large, most people were behind the change. An opinion survey showed that three-quarters of those asked were broadly in favor of legalizing euthanasia.[79] On May 16, 2002, the lower house of the Belgian parliament endorsed the bill by 86 votes in favor, 51 against and with 10 abstentions.[80]

73 Fact sheet based on a paper by Professor Meinrad Schaer (December 16, 1996), published by the South Australian Voluntary Euthanasia Society, *op. cit*. The data were verified by Derek Humphry (personal correspondence on January 4, 1999) who said that there has never been a prosecution for abuse of the law. Further information is available at: http://www.ofj.admin.ch/f/index.html; Exit/A D M D (Suisse Romande) C.P. 100, 1222 Vesenaz, Geneva. Telephone: 41-22-735-7760; FAX: 41-22-735-7765; EXIT/Vereinigung fr humanes Sterben (German-speaking) Feldeggstraase 3, P.O. Box 309, CH-8034 Zurich/Switzerland. Phone: +41-1-383 33 53; FAX: +41-1-383 33 78.

74 World Right-to-Die *Newsletter*, Issue 42 (November 2002), p. 7. Cf. www.worldrtd.org

75 Luc Deliens, Freddy Mortier, Johan Bilsen *et al.*, "End-of-life Decisions in Medical Practice in Flanders, Belgium: A Nationwide Survey," *Lancet,* Vol. 356 (November 25, 2000): 1806–1811.

76 Tony Sheldon, "Belgium Considers Legalizing Euthanasia," *British Medical Journal*, Vol. 320 (January 15, 2000): 137.

77 Luc Deliens, Freddy Mortier, Johan Bilsen *et al.*, "End-of-life Decisions in Medical Practice in Flanders, Belgium: A Nationwide Survey," *op. cit.*, at 1806. See also F. Mortier, L. Deliens, J. Bilsen, M. Cosyns, K. Ingels and R. Vander Stichele, "End-of-life Decisions of Physicians in the City of Hasselt (Flanders, Belgium)," *Bioethics*, Vol. 14 (July 2000): 254–267; Luc Deliens and Jan Bernheim, "Palliative Care and Euthanasia in Countries with a Law on Euthanasia," *Palliative Medicine*, Vol. 17 (2003): 393–394.

78 Wim Weber, "Belgian Euthanasia Bill Gains Momentum," *Lancet,* Vol. 357 (February 3, 2001), p. 370.

79 Andrew Osborn, "Belgians follow Dutch by legalising euthanasia," *The Guardian* (October 26, 2001).

80 See the website of the Belgian Senate: www.senate.be/home/home.html under the number 2-244/23.

The legislation lays out the terms for doctors to end the lives of terminally ill patients – though, with doctors operating an informal system of euthanasia to some extent, no immediate or radical changes are expected in the way they function in Belgium. Patients must be at least 18 years old and have made specific, voluntary and repeated requests that their lives be ended. They must put this in writing. (Research indicates that Flemish doctors frequently disregard this issue and do not obtain the patient's request to end his/her life).[81] Requests will be approved only if the patient is terminally ill, in constant suffering and of sound mind. At least one month must elapse between the written request, which can be made by a nominated adult if the patient is incapable of writing, and the mercy killing. An independent physician must be consulted. The consulted MD is required to inspect the medical file, to examine the patient and must ascertain the enduring and unbearable physical or mental suffering. The physician is then required to write a report. In a case of a nursing team that has a regular contact with the patient, the GP must discuss the request of the patient with that team or with members of that team. If the patient wishes so, the GP must discuss the request of the patient with the proxies indicated by the patient. Controversially, there is also provision for patients who are not in the final phases of a terminal illness to opt for euthanasia. Such a request requires a further authorization by a psychiatrist or a specialist in the disease. The consultant must be independent with regard to the patient, his/her GP, and also with regard to the first consultant.[82] All cases of mercy killing have to be fully documented in a special format and presented to a permanent monitoring committee, appointed by the government. If a case does not find the approval of at least two thirds of the committee members, which include 4 jurists and 12 physicians, 4 of them specialists for anesthesia and pain reduction, it is to be handed over to the state prosecution for further investigation.[83]

Belgian doctors voiced opposition to the new law because it opens the door for euthanasia too wide. Marc Moens said that "Doctors know that this law is simply flawed and find it totally unacceptable that individuals who are not terminally ill will also be eligible for euthanasia."[84]

For its part, English criminal law does not recognize active euthanasia as a defense in a murder charge. Lord Devlin directed the jury in *R. v. Adams* that no doctor, nor any man, no more in the case of the dying than of the healthy, has the right to deliberately cut the thread of life.[85] This argument was restated in *R. v. Cox.*[86] At the same time, it is established that passive euthanasia may be allowed in certain circumstances. Precedents prescribe

81. Freddy Mortier and Luc Deliens, "The Prospects of Effective Legal Control of Euthanasia in Belgium: Implications of Recent End of Life Studies," in A. Klijn, M. Otlowski and M. Trappenburg (eds.), *Regulating Physician-negotiated Death* (Gravenhage: Elsevier, 2001), p. 180.

82 Chapter II (Conditions and procedure), art. 3, § 2, 3 of law proposal no. 2-244/23. See website of the Belgian Senate: http://www.senate.be/home/home.html

83 *Rationalist International*, Bulletin 104 (October 27, 2002), http://www.rationalistinternational.net/

84 Justin Sparks, "Belgian docs unhappy about proposed euthanasia law," *Reuters Health* (May 16, 2002). See also Andrew Osborn, "Belgian outcry over first mercy killing under new law," *The Guardian* (October 9, 2002); Herman Nys, "Euthanasia in the Low Countries: A Comparative Analysis of the Law Regarding Euthanasia in Belgium and the Netherlands," *Ethical Perspectives*, Issue 9/2 (June 2002): 73–85.

85 *R. v. Adams*, *Criminal Law Rev.* (1957), at 365–377.

86 See "GMC Tempers Justice with Mercy in Cox Case," *British Medical Journal*, Vol. 305 (November 1992), p. 1311.

withholding medical care if such a course of action represents good medical practice,[87] and if it is done "in the best interests of the patient."[88] This reasoning, which accentuates the best interests of the patient, has been reiterated in several court decisions and has become a cornerstone in English law. When determining these interests, the court balances the benefit of continued treatment against the pain and suffering of the patient concerned. In *Re J.*, Lord Donaldson of the Court of Appeal delivered an opinion against resuscitating a severely brain-damaged child because the pain and suffering likely to be experienced exceeded any benefit accruing from prolonging his life. Lord Donaldson argued that there are cases "in which the answer must be that it is not in the interests of the child to subject it to treatment which will cause increased suffering and produce no commensurate benefit, giving the fullest possible weight to the child's, and mankind's desire to survive."[89]

All requests for withdrawal of tube feeding in England must go through the courts. In the *Airedale NHS Trust v. Bland* case involving a football fan who was severely injured in the Hillsborough stadium disaster of April 1989, both the Court of Appeal and the House of Lords upheld a declaratory judgment by the Family Division of the High Court that withdrawing artificial nutrition and hydration from a patient in a severe, persistent vegetative state (PVS)[90] did not constitute an unlawful act. The court held that artificial feeding and the administration of antibiotic drugs could lawfully be withheld from an insensate patient who had no hope of recovery when it was believed that the result would be that the patient would die shortly thereafter. It was also emphasized that by virtue of his condition, a PVS patient would not suffer as a result of being deprived of food and hydration, and that the major consideration in determining the best interests of the patient are medical: the opinion of the physicians decides the course of the treatment.[91] The court explained that relevant considerations for deciding the best interests of PVS patients include the avoidance of invasive and undignified procedures, which would have an adverse effect upon the way such patients would be remembered by their loved ones. Lord Goff saw no reason to maintain medical treatment simply to prolong a patient's life when such treatment had no therapeutic purpose of any kind, "as where it is futile because the patient is unconscious and there is no prospect of any improvement in his condition."[92] Lord Mustill argued that Anthony Bland had no best interests because the loss of all cognitive functions meant that he had "no best interests of any kind." Because the patient had no interest in staying alive, no legal justification existed for any invasive life-supporting treatment.[93]

In October 2000, the courts reiterated the same rationale while resorting for the first time to the *Human Rights Act* of 1998, which incorporates the European Convention on Human

87 *Bolam v. Friern Hospital Management Committee* (1957), 2 All ER 118.

88 *F v. West Bershire Health Authority* (1989) 2 All ER 545, at 546. See also *Re J* (1990) 3 All ER 930.

89 *Re J*, at 938.

90 I resent this term, thinking it is unethical and use instead the term Post-Coma Unawareness. See R. Cohen-Almagor, "Language and Reality at the End of Life," *Journal of Law, Medicine and Ethics*, Vol. 28, No. 3 (Fall 2000): 267–278.

91 *Airedale NHS Trust v. Bland* (1993) 1 All ER 821.

92 *Ibid.*

93 *Ibid*, at 894. For further discussion, see *Law Hospital NHS Trust v. Lord Advocate and Others*, Court of Session: Inner House (First Division) (22 March 1996), Inner House Cases; House of Lords, *Select Committee on Medical Ethics*, session 1993–94, Vol. II, Minutes of Oral Evidence (London: HMSO, 1994); Joan Loughrey, "Medical Decision Making and the Human Rights Act 1998," *Proceedings of the 13th World Congress on Medical Law* (Helsinki, August 6–10, 2000), Vol. II, pp. 687–695.

Rights into UK law. The courts ruled in the cases of *NHS Trust A v M* and *NHS Trust B v H* that it was lawful to withdraw artificial nutrition and hydration from two patients in PVS. Four consultants in one case, and five in the other, had diagnosed the medical condition – a diagnosis which the Official Solicitor had not challenged and which fell within the guidelines established by the Royal College of Physicians for the determination of PVS. Therefore, "the continuation of artificially supplying nutrition and hydration to both patients was not in their best interest and could be withdrawn subject to being lawful within the provisions of the European Convention on Human rights, as incorporated into the domestic law of England and Wales by the *Human Rights Act* 1998, on 2 October 2000." Article 2 of that Act protects the right to life, but the court held that existing practice in the United Kingdom, of withdrawing artificial nutrition and hydration, was in compliance with the provisions of Article 2 and was compatible with the values of democratic societies across the world.[94]

In August 2001, the High Court in London has granted a woman who suffers from the incurable degenerative disorder motor neurone disease (Amyotrophic Lateral Sclerosis) permission to seek judicial review of the Director of Public Prosecutions' (DPP) refusal to allow her husband of 25 years, Brian, to end her life. Diane Pretty, 47, was paralyzed from the neck down, had to be fed by tube, and had no decipherable speech, though her intellect was unimpaired. She claimed her quality of life had become so low that she had the right under human rights legislation to choose to end her life. She had exhausted all medical alternatives and had accessed palliative care services. However, knowing the inevitable progression of her disease, and the further distress it will cause her, she had decided that she wanted to die now. Ms. Pretty's plea was supported by the Voluntary Euthanasia Society and civil rights group Liberty. Liberty asked the DPP, David Calvert-Smith, to guarantee her husband would not be prosecuted for aiding and abetting a suicide under Section 2 of the Suicide Act if he tried to help her.[95]

In October 2001, Ms. Pretty argued her case in the Queen's Bench Divisional Court that the blanket legal ban on assisting a suicide denied her the right to "die with dignity" and breached the European Convention on Human Rights. Her QC, Philip Havers, had told the judges that she was close to death and "frightened and distressed" at the suffering and indignity she would have to suffer if the disease ran its course. Her condition meant she was physically unable to kill herself, hence she wanted her husband to help her when the time came. Lord Justice Tuckey, Lady Justice Hale and Mr. Justice Silber ruled that parliament had given the DPP no power to rule out a prosecution in advance, and that the prohibition on assisting someone to commit suicide did not contravene the European convention. Lord Justice Tuckey rightly said: "We are being asked to allow a family member to help a loved one die, in circumstances of which we know nothing, in a way of which we know nothing, and with no continuing scrutiny by any outside person."[96] Assisted suicide should not be carried by family members but rather by qualified physicians.

Ms. Pretty appealed to the highest court without much success. The unanimous decision by five law lords denied her the right to appeal a decision by the DPP. Lord Bingham said

[94] Chris Docker, Press Release from Liberty and The Voluntary Euthanasia Society (August 17, 2001), circulated by Derek Humphry via ergo@efn.org.

[95] "Head to Head: Assisted Suicide," and Court Rules for 'Right to Die' Woman," *British Broadcasting Corp. News Online* (Friday, August 31, 2001): http://news.bbc.co.uk/hi/english/uk

[96] Clare Dyer, "Dying wife loses 'Suicide Aid' Case," *The Guardian* (October 19, 2001); *The Times* (October 23, 2001).

the European Convention on Human Rights, enshrined in the UK *Human Rights Act*, did not guarantee assisted suicide, and that Ms. Pretty could not establish that her rights had been infringed by the DPP's refusal to waive any legal action against her husband. Lord Bingham rejected the argument that the right to life protected the right to self-determination over life and death. On the right of freedom of thought, Lord Bingham said that Ms. Pretty might have a sincere belief in the virtue of assisted suicide and was free to express that view. But it could not mean there was a requirement that her husband should be absolved from the consequences of conduct which was against the law.

Lord Bingham maintained that assisted suicide and consensual killing were unlawful in all Convention countries except The Netherlands, but even there Mr. Pretty would be liable if he were to assist Mrs. Pretty to take her own life. He rightly asserted that it would have been a gross dereliction of the DPP's duty and a gross abuse of his power had he ventured to undertake that a crime yet to be committed would not lead to prosecution. The claim against him had to fail on that ground alone.[97]

Immediately after the hearing Ms. Pretty said: "I want to go on. I feel I have no rights. The law lords don't want to admit that the law is wrong." The Prettys vowed to fight on in the European Court of Human Rights.[98] However, a panel of seven judges of the European Court of Human Rights ruled unanimously that Britain had not violated Europe's human rights convention by refusing to grant Brian Pretty immunity from prosecution if he helps his wife Diane commit suicide.[99]

The French Penal Code distinguishes between "active" and "passive" euthanasia. The former involves direct intervention to cause the death of a patient. The latter refers to the foregoing of life-sustaining treatment for terminally ill people. Active euthanasia is regarded as murder while passive euthanasia is considered an offence against a French law, which makes it a crime not to help a person in danger. Recent studies said that almost half the deaths recorded as taking place in intensive care units in France resulted from a decision to stop treatment and could be classed as acts of "passive" euthanasia. Edouard Ferrand and colleagues expressed concern that that there are no guidelines to govern and justify the withholding or withdrawal of life-saving treatments. They explained this by saying that the relationship between patient and physician is limited in France to a traditional paternalism, based on the principle of beneficence. Only 42% of decisions were notified in the medical record, which may reflect the reluctance of physicians to record their decisions in the French legal circumstances.[100]

97 *Regina (Pretty) v Director of Public Prosecutions, Secretary of State for the Home Department,* Before Lord Bingham of Cornhill, Lord Steyn, Lord Hope of Craighead, Lord Hobhouse of Woodborough and Lord Scott of Foscote (Speeches November 29, 2001); Law Report, "No breach in proleptic refusal to grant immunity," *The Times* (December 5, 2001).
See http://www.thetimes.co.uk/article/0,,12-2001561814,00.html

98 "Diane Pretty loses right to die appeal," *The Guardian* (November 29, 2001).

99 Paul Ames, "European court rejects British woman's assisted suicide appeal; supporters pledge to continue right to die campaign," *Associated Press* (April 29, 2002). For further discussion, see G.E. Dickinson, C.J. Lancaster, D. Clark, S.H. Ahmedzai and W. Noble, "UK Physicians' Attitudes toward Active Voluntary Euthanasia and Physician-assisted Suicide," *Death Studies*, Vol. 26 (July–August 2002): 479–490.

100 Edouard Ferrand, René Robert, Pierre Ingrand, François Lemaire, "Withholding and Withdrawal of Life Support in Intensive-care Units in France," *The Lancet*, Vol. 357, No. 9249 (January 6, 2001). http://www.thelancet.com/

In 2000, France's National Ethics Committee has said that euthanasia may be allowed in certain circumstances. But the Committee underscored that this does not mean euthanasia should be decriminalized. In a report, which took three years to prepare, the committee speaks of the need for compassion where therapy has failed and when patients ask to be relieved of unbearable suffering. "If there is no other solution, if palliative care and pain-killers are ineffective, if all treatment or therapy has failed, if there is unanimous agreement that the situation has become intolerable, then one can envisage euthanasia," said Ethics Committee President Dr Didier Sicard.[101] It marks a turnaround in the committee's thinking and its first recommendations on euthanasia for nine years. In 1991, it rejected a European Parliament proposal that euthanasia be carried out in hospitals and care centres. One committee member had estimated that there are about 2,000 clandestine acts of assisted suicide in France each year.[102]

In July 2001, the French Health Minister, Bernard Kouchner, has admitted he practiced euthanasia during his career when he served as a doctor in the war zones of Vietnam and Lebanon. Mr Kouchner – a founding member of the Paris-based medical aid agency, Medecins sans Frontieres (MSF) (Doctors without Frontiers) – also acknowledged that passive euthanasia, where doctors suspend treatment of dying patients, occurs frequently in France. However, although opinion polls showed wide support for the euthanasia practice in certain cases,[103] Mr Kouchner said he had no plans to legalize euthanasia.[104]

In Canada, parliament debated a private member's bill (C-261) to legalize active euthanasia. The bill was not adopted, but legislators became aware that physician-assisted suicide had widespread support in Canada. About three quarters of Canadians (77%) believe that doctors should be allowed to end the life of a patient whose life is immediately threatened by a disease that causes the patient great suffering. Canadians are less likely to support physician-assisted suicide if the patient is suffering from a disease that is not immediately life threatening, such as a chronically debilitating illness. Still, 57% of Canadians believe that doctors should be allowed, by law, to end the life of a patient who suffers from a disease that does not immediately threaten his or her life. This figure has not changed since 1995.[105]

Two of the Canadian court cases should be mentioned. One concerns Nancy B., a twenty-five-year-old woman who had had generalized polyneuropathy for two and a half years as a result of Guillain-Barre disease. She initiated a legal action for an injunction permitting her

[101] The committee only has consultative powers, but most of its recommendations to French lawmakers are heeded. It has made recommendations on sensitive topics such as the sterilization of the mentally handicapped and obligatory psychiatric care for rapists.

[102] BBC London report, "2,000 assisted suicide in France," Right To Die list <right_to_die@efn.org> http://www.FinalExit.org ; see also website of Association pour le Droit de Mourir dans la Dignite, URL: http://perso.infonie.be/admd

[103] Polls conducted in 1987 and 1997 showed that wide majority (85% in 1987 and 84% in 1997) were in favor of granting terminal patients in unbearable suffering the right to receive assistance in dying on request. 63% in 1987 and 55% in 1997 were in favor of active euthanasia. Cf. website of Association pour le Droit de Mourir dans la Dignite (ADMD), URL: http://perso.club-internet.fr/admd

[104] BBC: French Minister Admits Mercy Killings (July 24, 2001): http://news.bbc.co.uk; *Reuters*, "France has no plans to legalize euthanasia" (July 25, 2001). See also Craig S. Smith, "Son's wish to die, and mother's help, stir French debate," *New York Times* (September 27, 2003).

[105] Gary Edwards and Josephine Mazzuca, "Three Quarters of Canadians Support Doctor-Assisted Suicide," *Gallup News Service* (March 24, 1999), http://www.gallup.com/poll/releases/pr990323.asp. See also Frederick H. Lowy *et al.*, *Canadian Physicians and Euthanasia* (Ottawa: Canadian Medical Association, 1993), p. 3.

physicians to withdraw the respirator. The Quebec Superior Court granted the injunction, her respirator was withdrawn and Nancy died.[106] In Justice Dufour's opinion, Nancy B.'s refusal of treatment was not an attempt to commit suicide but rather an attempt merely to allow a disease to take its natural course.

The most known death-with-dignity case in Canada concerns Sue Rodriguez who was dying from Amyotrophic Lateral Sclerosis (ALS), a disease that causes progressive paralysis of the muscles: muscles of the face, the tongue, the throat, the respiratory system, the shoulders, hands and legs. In its final stages the patient cannot swallow, speak, cough, or use his respiratory muscles.[107] One specialist described this situation as "a living hell."[108] As her condition deteriorated, Ms. Rodriguez publicly expressed a wish to have a physician assist her in ending her life at a time of her choosing, when she herself would be unable to do so, rather than waiting helplessly to die by suffocation or choking. Ms. Rodriguez sought to challenge the Criminal Code of Canada prohibition on assisted suicide, on the grounds that it violated the *Charter of Rights and Freedoms*. The specific section of the Criminal Code is 241(b): "Everyone who aids or abets a person to commit suicide, whether suicide ensues or not, is guilty of an indictable offence and liable to imprisonment for a term not exceeding fourteen years."[109]

The appeal was rejected in a 5 (Sopinka, La Forest, Gonthier, Iacobucci and Major) to 4 (McLachlin, L'Heureaux-Dube, Lamer and Cory) landmark decision. The Court did not want to intervene in this delicate public matter, thinking that it is up to the legislature to change the law if such a change was deemed necessary.[110] I discussed her case with three of the justices in the Canadian Supreme Court. One of them told me that this was the toughest decision s/he has ever made, and that the Court might overturn the decision if the legislature failed to address the issue adequately and another case came up.[111]

Patients in a devastating situation, who wish to cease living, if helped by a physician, are helped to relieve their suffering. The motivation is to assist one's fellow by providing relief from enduring suffering. The decision to perform physician-assisted suicide is first and foremost a *moral* decision. The physician who provides the assistance is convinced that this

[106] *Nancy B. v. Hotel-Dieu de Quebec et al.* (1992), 86 DLR (4th) 385 (Que Sup Ct). For criticism of this ruling, see Arthur Fish and Peter A. Singer, "Nancy B.: The Criminal Code and Decisions to Forgo Life-sustaining Treatment," *Can. Med. Ass. J.*, Vol. 147 (September 1992): 637–642; Bernard M. Dickens, "Medically Assisted Death: *Nancy B. v. Hotel-Dieu de Quebec*," *McGill Law Journal*, Vol. 38, No. 4 (October 1993): 1053–1070.

[107] For general discussions concerning the progressive neuro-muscular disease Amyotrophic Lateral Sclerosis (ALS) and assisted suicide, see the October 1, 1998 issue of the *New England Journal of Medicine*. URL: http://www.nejm.org

[108] (Tel Aviv) 1141/90 *Benjamin Eyal v. Dr. Nachman Willensky and Others*, Vol. 51, No. 3, pp. 187, 192.

[109] R.S.C. 1985, c. C-46.

[110] *Rodriguez v. British Columbia (Attorney General)*, [1993] 3 S.C.R. 519. For a critical discussion, see Lorraine Eisenstat Weinrib, "The Body and the Body Politic: Assisted Suicide under the *Canadian Charter of Rights and Freedoms*," *McGill Law Journal*, Vol. 39 (1994): 619–644; Jerome E. Bickenbach, "Disability and Life-Ending Decisions," in Margaret P. Battin, Rosamond Rhodes and Anita Silvers (eds.), *Physician Assisted Suicide* (New York and London: Routledge, 1998): 123–132.

[111] Personal discussion at the Canadian Supreme Court, Ottawa (September 28, 1998). In 1997, Robert Latimer was convicted of second-degree murder for the mercy killing of his severely disabled daughter. He is said to prepare an appeal to the Supreme Court of Canada to overturn his sentencing. Cf. Barney Sneiderman, "The Case of Robert Latimer: A Commentary on Crime and Punishment," *Alberta L. Rev.*, Vol. 37, No. 4 (1999): 1017–1044; *idem*, "Latimer in the Supreme Court: Necessity, Compassionate Homicide, and Mandatory Sentencing," *Saskatchewan L. Rev.*, Vol. 64, No. 2 (2001): 511–544. See also articles in *Criminal Reports*, Vol. 39, Fifth Series (April 2001).

act is justified not only medically but also morally, otherwise he or she would not have agreed to assist the patient in the first place.

CONCLUSION

This chapter surveys the attempts that have been made around the globe to facilitate "death-with-dignity." In the United States, ten states during the past five years have passed bills making euthanasia or PAS illegal, and bills are pending in five more.[112] Oregon's Measure 16 that allows assisted suicide is facing a challenge. In Australia, the Northern Territory Act that allowed terminally ill patients to commit suicide with a doctor's help was declared void. The legislatures of England and Canada resist attempts to legalize assisted suicide and euthanasia. Switzerland condones the practice of assisted suicide but not of euthanasia. The Netherlands and Belgium are the only countries in the liberal world that accept the policy and practice of both euthanasia and physician-assisted suicide, without seeing much difference between the two. The legalization process of euthanasia in the Netherlands and Belgium during the years 2001–2002 may give fresh impetus to campaigns for legal mercy killing elsewhere in Europe – especially in Britain and France, where significant movements are pressing for it.[113]

[112] Ezekiel J. Emanuel, "Euthanasia: Where the Netherlands Leads Will the World Follow?," *British Medical Journal,* Vol. 322 (June 9, 2001): 1376. For a general overview about assisted suicide laws around the world, see http://www.assistedsuicide.org/suicide_laws.html

[113] For further discussion, see Brian J. Pollard, "Can Euthanasia Be Safely Legalized?," *Palliative Medicine*, Vol. 15 (January 2001): 61–65; Peter Singer, "Voluntary Euthanasia: A Utilitarian Perspective," *Bioethics*, Vol. 17, Nos. 5-6 (2003):526-541.

Part A

Background

CHAPTER 1

THE THREE RESEARCH REPORTS OF 1990, 1995 AND 2001, AND THEIR INTERPRETATIONS

1. PRELIMINARIES

The Dutch understanding of euthanasia is said to be marked by its precision. Unlike other countries that distinguish between active and passive euthanasia, between direct and indirect euthanasia, and between voluntary and involuntary euthanasia, the Dutch definition of the term is exact: the intentional taking of someone's life at his or her explicit request. According to the law, only a competent patient's request can be accepted. Consequently:

a. What is termed "euthanasia" in the Netherlands is called "active euthanasia" in other parts of the world. In the Dutch conception, euthanasia is active by definition and there is no need to specify the act by the term "active," as other countries do.

b. All other kinds of end-of-life ("terminal") care bear other names. Thus, for instance, withdrawal of treatment is not considered euthanasia. Elsewhere it is termed "passive euthanasia." In the Netherlands, this term is deemed illogical and useless insofar as passive euthanasia is self-contradictory because it concerns the omission of a treatment to which the patient has not consented.[1] Within the law, the difference between acting and refraining from acting has no particular relevance, and such a legal consideration takes precedence over the psychological experience of the difference. The prevailing Dutch perspective regards any treatment that has no clear medical benefit for the patient as futile. The argument is that no patient should be subjected to useless medical interventions, or that these treatments should be offered as an option to patients. In a leading test case, a Dutch court ruled that contrary to the wishes of the next of kin, a hospital was not obliged to return an 80-year-old cancer patient to intensive care when his condition was deteriorating.[2]

c. Unintended shortening of life in the course of treatment to abate suffering is viewed in the Dutch discussion as a physician's duty to alleviate pain. Some call this "indirect euthanasia," and others see it as coming within the "double effect doctrine."[3] The possibility

[1] G.K. Kimsma and E. van Leeuwen, "Dutch Euthanasia: Background, Practice, and Present Justifications," *Cambridge Q. of Healthcare Ethics*, Vol. 2 (1993), p. 24.

[2] Cf. Tony Sheldon, "Doctors Not Obliged to Carry Out Treatment They Think 'Futile'," *British Medical Journal*, Vol. 319 (October 23, 1999), at 1088.

[3] The ethical concept of double effect is used to justify medical treatment designed to relieve suffering where death is an unintended, though foreseeable, consequence. The doctrine is based on two basic presuppositions: (1) the doctor's intention is to alleviate suffering; (2) the treatment must be proportional to the illness. The doctrine applies if the desired outcome is judged to be "good" (e.g. relief of suffering); the "bad" outcome (e.g. death of patient) is not intended; the "good" outcome is not achieved by means of the "bad," and the "good" outcome outweighs the "bad." For further discussion on the double effect doctrine, see R. Cohen-Almagor, "Language and Reality at the End of Life," *The Journal of Law, Medicine and Ethics*, Vol. 28, No. 3 (Fall 2000): 267–278; Daniel P. Sulmasy and Edmund D. Pellegrino, "The Rule of Double Effect," *Archives of*

that a seriously ill patient might die as a result of such treatment does not constitute euthanasia.

d. The decision to end a life without an explicit request is legally termed "killing a person" rather than euthanasia, although some physicians would disagree. Involuntary euthanasia is regarded as a contradiction in terms.

e. The distinction between euthanasia and physician-assisted suicide (PAS) hardly figures in the Dutch discussion. Although assisted suicide was considered for many years to be a crime by law and the distinction may be seen as morally relevant, no relevance had been attributed to this distinction in the actual medical context in the cases of euthanasia that had been brought to trial. In both practices, the physician had to meet the same substantive and procedural requirements.[4] Many of my interviewees said that euthanasia and physician-assisted suicide were and are considered one and the same in the Netherlands, implying the same responsibility of the physician.

Since November 1990, prosecution is unlikely if a doctor complies with the Guidelines set out in the non-prosecution agreement between the Dutch Ministry of Justice and the Royal Dutch Medical Association. These Guidelines are based on the criteria established in court decisions relating to the conditions under which a doctor can successfully invoke the defense of necessity. The substantive requirements are as follows:

(1) The request for euthanasia or physician-assisted suicide must be made by the patient and must be free and voluntary.

(2) The patient's request must be well considered, durable and consistent.

(3) The patient's situation must entail unbearable suffering with no prospect of improvement and no alternative to end the suffering.[5] The patient need not be terminally ill to satisfy this requirement and the suffering need not necessarily be physical.

(4) Euthanasia must be a last resort.[6]

Internal Medicine, Vol. 159 (March 22, 1999): 545–550; Len Doyal, "The Moral Character of Clinicians or the Best Interests of Patients?," *British Medical Journal*, Vol. 318 (May 29, 1999): 1432–1433.

4 Cf. Johan Legemaate, "Twenty-Five Years of Dutch Experience and Policy on Euthanasia and Assisted Suicide: An Overview," in David C. Thomasma, Thomasine Kimbrough-Kushner, Gerrit K. Kimsma, and Chris Ciesielski-Carlucci (eds.), *Asking to Die* (Dordrecht: Kluwer Academic Publishers, 1998), p. 20; Gerrit K. Kimsma, "Euthanasia and Physician Assisted Suicide in the Netherlands," in *Medizin, Ethik, Recht* (1994): 161–169.

5 The Medical Association Executive Board emphasized that there are only limited possibilities for verifying whether suffering is unbearable and without prospect of improvement. In any case, the Board considered it the doctor's task to investigate whether there are medical or social alternatives that could make the patient's suffering bearable. John Griffiths, Alex Bood and Heleen Weyers, *Euthanasia and Law in the Netherlands* (Amsterdam: Amsterdam University Press, 1998), p. 66.

6 John Keown, "The Law and Practice of Euthanasia in the Netherlands," *The Law Quarterly Review*, Vol. 108 (January 1992), p. 56.

The procedural requirements are as follows:

(1) No doctor is required to perform euthanasia, but those opposed on principle must make this position known to the patient early on and help the patient to get in touch with a colleague who has no such moral objections.

(2) Doctors taking part in euthanasia should preferably and whenever possible have patients administer the fatal drug themselves, rather than have a doctor apply an injection or intravenous drip.[7]

(3) A doctor must perform the euthanasia.

(4) Before the doctor assists the patient, the doctor must consult a second independent doctor who has no professional or family relationship with either the patient or doctor. Since the 1991 *Chabot* case,[8] patients with a psychiatric disorder must be examined by at least two other doctors, one of whom must be a psychiatrist.

(5) The doctor must keep a full written record of the case.

(6) The death must be reported to the prosecutorial authorities as a case of euthanasia or physician-assisted suicide and not as a case of death by natural causes.[9] Since the legalization of the new law, cases of euthanasia and PAS are reported to the regional committees instead of the prosecutorial authorities.

2. THE THREE RESEARCH REPORTS

In 1990, the Dutch government appointed a commission to investigate the medical practice of euthanasia. The Commission, headed by Professor Jan Remmelink, Solicitor General to the Supreme Court, was asked to conduct a comprehensive nation-wide study of "medical decisions concerning the end of life (MDEL)." The following broad forms of MDEL were studied:

(1) Non-treatment decisions: withholding or withdrawing treatment when treatment would probably have prolonged life;

(2) Alleviation of pain and symptoms: administering opiates in such dosages that the patient's life might be shortened;

(3) Euthanasia and related MDEL: the prescription, supply or administration of drugs with the explicit intention of shortening life, including euthanasia at the patient's request, assisted suicide, and life termination without explicit and persistent request.[10]

7 The Royal Dutch Medical Association's refinements of the 1984 Guidelines (August 25, 1995). Cf. Marlise Simons, "Dutch doctors to tighten rules on mercy killings," *The New York Times* (September 11, 1995), p. A3.

8 Supreme Court of the Netherlands, Criminal Chamber (June 21, 1994), no. 96.972. For translation, see John Griffiths, Alex Bood and Heleen Weyers, *Euthanasia and Law in the Netherlands, op. cit.,* Appendix II (2), pp. 329–340.

9 Marcia Angell's Editorial, "Euthanasia in the Netherlands – Good News or Bad?," *New Eng. J. of Medicine,* Vol. 335, No. 22 (November 28, 1996); Adriaan Jacobovits, "Euthanasia in the Netherlands," *Washington Post* (January 23, 1997), p. A16; General Health Council, "A Proposal of Advice Concerning Careful Requirements in the Performance of Euthanasia" (The Hague, 1987).

10 Cf. P.J. van der Maas, J.J.M. van Delden, and L. Pijnenborg, *Euthanasia and other Medical Decisions Concerning the End of Life*, Health Policy Monographs (Amsterdam: Elsevier, 1992).

The study was repeated in 1995 and in 2001, making it possible to assess whether there were harmful effects over time that might have been caused by the availability of voluntary euthanasia in the Netherlands.[11] It is still difficult to make valid comparisons with other countries because of legal and cultural differences, and also because similar comprehensive studies are quite rare.[12]

The three Dutch studies were said to give the best estimate of all forms of MDEL (*i.e.*, all treatment decisions with the possibility of shortening life) in the Netherlands as approximately 39% of all deaths in 1990, 43% in 1995, and 44% in 2001. In the third category of MDEL, the studies gave the best estimate of voluntary euthanasia as 2300 persons each year (1.9% of all deaths) in 1990[13] and 3250 persons each year (2.4%) in 1995. In 2001, the rate of euthanasia increased to 2.6%. The estimate for physician-assisted suicide was about 0.3% in 1990, 1995 and 2001. There were 8900 explicit requests for euthanasia or assisted suicide in the Netherlands in 1990, 9700 in 1995, and 9700 in 2001.[14] Less than 40% were actually undertaken. Euthanasia and PAS were mainly performed among patients dying of cancer. The most worrisome data are related to the hastening of death without the explicit request of patients. There were 1000 cases (0.8%) without explicit and persistent request in 1990, and 900 such cases (0.7%) in 1995.[15] This number remained unchanged in 2001. Ending of life without a patient's explicit request occurred most frequently among people dying at age younger than 65 years.[16] There were still treatment alternatives in 8% of cases in which a life-terminating act was performed without explicit request of the patient. The physician did not use these alternatives when the patient indicated a desire to stop treatment

11 Paul J. van der Maas, Gerrit van der Wal, Ilinka Haverkate *et al,* "Euthanasia, Physician-Assisted Suicide, and Other Medical Practices Involving the End of Life in the Netherlands, 1990–1995," *New Eng. J. of Med.,* Vol. 335, No. 22 (November 28, 1996): 1699–1705; B.D. Onwuteaka-Philipsen, A. van der Heide, D. Koper, I. Keij-Deerenberg, J.A.C. Rietjens, M.L. Rurup, A.M. Vrakking, J.J. Georges, M.T. Muller, G. van der Wal, P.J. van der Maas, "Euthanasia and Other End-of-life Decisions in the Netherlands in 1990, 1995, and 2001," *Lancet*, Vol. 362 (August 2, 2003): 395–399.

12 For further discussion, see Johannes J.M. van Delden *et al.*, "Deciding Not to Resuscitate in Dutch Hospitals," *J. of Medical Ethics*, Vol. 19 (1993): 200–205; Tony Sheldon, "Euthanasia Law Does Not End Debate in the Netherlands," *BMJ*, Vol. 307 (December 11, 1993): 1511–1512; Henk Jochemsen, "Euthanasia in Holland: An Ethical Critique of the New Law," *J. of Medical Ethics*, Vol. 20 (1994): 212–217; Chris Ciesielski-Carlucci and Gerrit Kimsma, "The Impact of Reporting Cases of Euthanasia in Holland: A Patient and Family Perspective," *Bioethics*, Vol. 8, No. 2 (1994): 151–158; J.K.M. Gevers, "Physician Assisted Suicide: New Developments in the Netherlands," *Bioethics*, Vol. 9, No. 3/4 (1995): 309–312.

13 P.J. van der Maas, J.J.M. van Delden, and L. Pijnenborg, *Euthanasia and other Medical Decisions Concerning the End of Life*, *op. cit.*, p. 41.

14 B.D. Onwuteaka-Philipsen *et al.*, "Euthanasia and Other End-of-life Decisions in the Netherlands in 1990, 1995, and 2001," *op. cit.*, p. 396.

15 Gerrit van der Wal and Paul J. van der Maas, "Empirical Research on Euthanasia and Other Medical End-of-Life Decisions and the Euthanasia Notification Procedure," in David C. Thomasma, Thomasine Kimbrough-Kushner, Gerrit K. Kimsma, and Chris Ciesielski-Carlucci (eds.), *Asking to Die, op. cit.*, p. 171. See also Bill Mettyear, "advocating legalising voluntary euthanasia" (February 1997), http://www.on.net/clients/saves/ South Australian Voluntary Euthanasia Society. In his comments on the first draft of this study, van der Maas wrote that in 1990 the decision had been discussed with a patient in 46% of the cases and in 14% there had been an expressed wish. Because explicit request is defined very strictly in the Dutch studies, these were not counted as euthanasia on request. Van der Maas noted an interesting comparison: Replication studies in Australia and Belgium both found frequencies over 3% for ending of life without explicit request. He estimated the number of active cases involving ending of life among newborns in the Netherlands to be 10–15 cases per year. Personal communication on September 18, 2000.

16 B.D. Onwuteaka-Philipsen *et al.*, "Euthanasia and Other End-of-life Decisions in the Netherlands in 1990, 1995, and 2001," *op. cit.*, p. 397.

because it "only would prolong suffering," or because the expected gain was not enough to make the treatment worthwhile.[17]

In 1990, 30% of the general practitioners (GPs) interviewed said that they had performed a life-terminating act at some time without explicit request (as compared with 25% of specialists and 10% of nursing home physicians).[18] The numbers rose substantially in the two later reports. In 1995, 63% of the family physicians had ever performed euthanasia or PAS. In 2001, 71% performed those conducts. Among nursing home physicians the percentages were 21 in 1995, and 36 in 2001. The percentage of clinical specialists was 37 in both 1995 and 2001. For physicians who would never perform euthanasia, the proportion fell consistently: 4% in 1990, 3% in 1995, and 1% in 2001.[19] Physicians who oppose the practice of euthanasia are becoming extinct. Apparently, there are explicit and implicit pressures on physicians to accept the practice as part of the medical profession.

The proportion of physicians who were ever engaged in life-terminating acts without explicit request of the patients decreased from 27% in 1990 to 23% in 1995, and further to 13% in 2001.[20] It should be noted that the level of consultation was significantly lower in life-termination acts without patient's explicit request than in cases of euthanasia or physician-assisted suicide. A colleague was consulted in 48% of the cases (as compared with 84% in euthanasia and assisted suicide cases). Relatives were consulted in 72% of the cases (as compared with 94% in euthanasia and assisted suicide cases). In 68% of the cases, the physician felt no need for consultation because the situation was clear.[21] Van der Maas and colleagues note that this should be considered in light of the very brief period by which life was shortened.[22] In 67% of the cases, life was shortened by fewer than 24 hours. In 21% of the cases, life was shortened by up to one week.[23]

About a quarter of the troublesome category 1000 patients had expressed a wish for voluntary euthanasia previously.[24] The patient was no longer competent in almost all of those cases, and death was hastened by a few hours or days. A question then arises: if death was impending within hours or a few days, what's the rush? Why not wait, provided that with the right palliative care it is possible to keep the patient comfortable? A small number of cases (approximately 15) involved babies who were suffering from a serious congenital disorder and were barely viable; hence the doctor's decision, in consultation with the parents, to

[17] P.J. van der Maas, J.J.M. van Delden, and L. Pijnenborg, *Euthanasia and other Medical Decisions Concerning the End of Life*, *op. cit.*, p. 62.

[18] *Ibid.*, p. 58.

[19] B.D. Onwuteaka-Philipsen *et al.*, "Euthanasia and Other End-of-life Decisions in the Netherlands in 1990, 1995, and 2001," *op. cit.*, p. 397.

[20] *Ibid.*

[21] In another study among family doctors, one quarter of the physicians said that they did not ask for a second opinion before administering euthanasia or assisted suicide, and 12% of the GPs had no consultation with any professional health worker. Cf. G. van der Wal, J.Th.M. van Eijk, H.J.J. Leenen and C. Spreeuwenberg, "Euthanasia and Assisted Suicide. II. Do Dutch Family Doctors Act Prudently?," *Family Practice*, Vol. 9, No. 2 (1992), p. 140.

[22] P.J. van der Maas, J.J.M. van Delden, and L. Pijnenborg, *Euthanasia and other Medical Decisions Concerning the End of Life*, *op. cit.*, p. 65.

[23] *Ibid*, p. 66.

[24] Henk A.M.J. ten Have, "Euthanasia: The Dutch Experience," *Annals de la Real Academia Nacional de Medicina*, Tomo CXII (Madrid, 1995), p. 429.

hasten the end of life.[25] A pertinent question here relates to the accuracy of this estimation. It seems that a further investigation is required to ascertain that this data is correct.

The Remmelink Commission regarded these cases of involuntary termination of life as "providing assistance to the dying." They were justified because the patients' suffering was unbearable, standard medical practice failed to help and, in any event, death would have occurred within a week. The Commission added that actively ending life when the vital functions have started failing is indisputably normal medical practice: "It deserves recommendation that the reporting procedures in place… will in the future also cover the active termination of life by a doctor in the framework of help-in-dying without an explicit request by the patient," except in situations where there is "the beginning of irreversible, interrelated failure of vital functions." In this last case, "natural death would very quickly occur even if the doctor did not actively intervene..." The recommendation goes on to say that this is not the case with patients whose vital functions are still intact and who are subject to life-shortening treatment without explicit request. Such cases should be reported.[26]

The Dutch authorities acknowledge that not every case of termination of life is reported. This does not mean, however, that some doctors report cases and others fail to do so. The distinction is related, so it is claimed, to the nature of the case. Cases in which a patient's life has been terminated without his or her explicit request are usually not reported. The explanation provided for this alarming phenomenon is that doctors may be confronted with appalling suffering on the part of "terminally ill patients who are no longer able to make their wishes known."[27] These patients have no prospect of recovery and are no longer able to make their wishes known due to the failure of their body functions. In such circumstances, the doctor, in consultation with the patient's relatives, may decide to actively hasten death. It is claimed that approximately 50% of these patients have clearly made it known at an earlier stage that they would wish to die upon reaching such a condition.[28] According to Henk ten Have, only 25% have made an earlier statement about shortening their lives.

Henk Ten Have finds it significant that only 17% of the physicians involved in the 1000 cases under examination cited request by the patient as the reason to terminate life. This indicates an important shift in moral justification, as respect for autonomy had always been the prime argument in favor of euthanasia, publicly defended by physicians, lawyers and the courts. Here it appears not to be autonomy that is respected by the physicians, but other concerns, such as the patients' suffering, medicine's failure to heal, and the knowledge that patients will soon die.[29]

In effect, what we are witnessing is a two-tier strategy on the part of euthanasia advocates: Whereas with competent patients, they champion autonomy and the need to

[25] See 1996 Study Findings, "Euthanasia and other decisions concerning the end of life in the Netherlands," Foreign Information Department, Netherlands Ministry of Foreign Affairs.

[26] Remmelink Commission, *Rapport Medische Beslissingen Rond het Levenseinde* (The Hague: SDU, 1991), p. 37. See also Henk A.M.J. ten Have, "Euthanasia: The Dutch Experience," *Annals de la Real Academia Nacional de Medicina,* Tomo CXII (Madrid, 1995), p. 429. In his comments on the first draft of this study, Leenen wrote that the proposal of the Remmelink Commission was rejected by nearly all the Dutch commentators and also by the government. Letter dated July 25, 2000.

[27] 1996 Study Findings, "Euthanasia and other decisions concerning the end of life in the Netherlands," Foreign Information Department, Netherlands Ministry of Foreign Affairs.

[28] *Ibid.*

[29] Henk A.M.J. ten Have, "Euthanasia: The Dutch Experience," *Annals de la Real Academia Nacional de Medicina,* Tomo CXII (Madrid, 1995): 429–430.

respect their wishes, with incompetent patients they emphasize the need to rely on medical judgment and to relieve patients from what is conceived to be unbearable suffering. Beneficence thus appears to be the major consideration in the Dutch euthanasia policy. The argument here is that when a doctor grants a patient's request for mercy killing, the basic motive is not respect for the patient's autonomy, but rather compassion. Accordingly, the doctor wants first and foremost to spare the patient misery and pain.[30]

On the basis of the 1995 report, the government decided to decrease the influence of criminal law in cases of euthanasia by instituting regional review committees. These committees were charged with reviewing each case of euthanasia reported to the medical examiner and advising whether to dismiss the case or to prosecute the physician involved. By introducing this mechanism, the government attempted to raise the willingness of physicians to report cases of euthanasia.

3. CONTRASTING INTERPRETATIONS

The literature on the Dutch policy and practice of euthanasia is vast, and opinions contradict one another considerably. Gerrit van der Wal and Robert Dillmann present a positive, almost ideal, portrayal of the situation, contending an "uphill" change of events. They claim that there has been a substantial increase in reported cases; a heightened awareness of the requirements of careful practice, with the necessary consideration given to patients' wishes; and a significant decrease in the proportion of physicians' prosecutions to almost nil. Most euthanasia cases take place at home among patients with a life expectancy of less than a month, after hospital treatment has proved ineffective, and the number of euthanasia cases reported in nursing homes is very low.[31]

Furthermore, the Dutch practice of euthanasia is not fueled by a scarcity of health care resources. Quite the opposite: The health care system has some features that allow euthanasia and other decisions at the end of life to be free from restraint and coercion. There is a "strongly developed system of primary care" and additional nursing care at home, as well as a "well developed system of care in nursing homes." The hospital system and specialist medical care "are also of a uniformly high quality." The nursing staff has a "well developed professional attitude and background and is usually involved in the decision making process."[32] Almost all patients have health care insurance; consequently, there are no financial incentives for hospitals, physicians, or family members to stop the care of patients. The legal right of patients to health care on the basis of their insurance "will override budget and other financial agreements."[33]

30 See G.A. den Hartogh, "Self-determination and Compassion in the Dutch Euthanasia-debate," *Rekenschap,* Vol. 39, No. 2 (1992), p. 110.

31 Gerrit van der Wal and Robert J.M. Dillmann, "Euthanasia in the Netherlands," *British Medical Journal,* Vol. 308 (1994): 1346–1349.

32 Arie J.G. van der Arend provides data that refute this contention. Cf. "An Ethical Perspective on Euthanasia and Assisted Suicide in the Netherlands from a Nursing Point of View," *Nursing Ethics,* Vol. 5, No. 4 (1998), at 313. See also *idem,* "Euthanasia and Assisted Suicide in the Netherlands: Clarifying the Practice and the Nurse's Role," *Int. Nurs. Rev.,* Vol. 45, No. 5 (1998), esp. p. 148; Ada van de Scheur and Arie van der Arend, "The Role of Nurses in Euthanasia: A Dutch Study," *Nursing Ethics,* Vol. 5, No. 6 (1998), esp. pp. 505–506.

33 Gerrit van der Wal and Robert J.M. Dillmann, "Euthanasia in the Netherlands," *op. cit.*: 1346–1349.

Margaret Battin emphasizes that euthanasia in the Netherlands is comparatively rare. Her interpretation of the Remmelink report is quite favorable. She points out that each year, less than one-third of the 9,000 explicit requests for euthanasia are actually honored. She also finds it positive that the practice of euthanasia is based on general practitioners insofar as it is typically performed by the patient's personal physician, rather than a stranger. Euthanasia is performed within the context of an extended period of consultation and care. Not only is it usually performed at home with the patient's family present, but the physician also remains with the patient or in an adjoining room throughout the process.[34]

Other advocates of the Dutch policy and practice of euthanasia reiterate that the findings of the 1995 report do not support the claim that the Dutch are on a slippery slope. A number of factors have contributed to the increase in voluntary euthanasia and medically assisted suicide from 2.1% to 2.7% of total deaths in the five-year period following the first report. They explain that mortality rates have risen as the population ages, with an increase in the proportion of deaths from cancer and heart disease. Life-prolonging techniques have become increasingly available, and there may have been generational and cultural changes in patients' attitudes. The slightly fewer cases of ending life without an explicit request may be an indication of the increasing openness with which end-of-life decisions are discussed with patients. The South Australian Voluntary Euthanasia Society responded to the results of the 1995 Dutch study by saying:

> Since the notification procedure was introduced, end-of-life decision making in the Netherlands has changed only slightly, in an anticipated direction. Close monitoring of such decisions is possible, and we found no signs of an unacceptable increase in the number of decisions or of less careful decision-making.[35]

Here again, however, the interpretations are inconsistent. Hendin and colleagues argue that by 1995, the number of deaths in which physicians gave pain medication with the explicit intention of ending the patient's life had increased from 1350 cases (1% of all deaths) to 1896 (1.4% of all deaths). They note that van der Maas and colleagues do not make these comparisons, maintaining that the total of all the deaths resulting from euthanasia, assisted suicide, ending life without consent, and giving opiates with the explicit intention of ending life must be considered. By so doing, it can be seen that the estimated number of deaths caused by active intervention of physicians has increased from 4813 (3.7% of all deaths) in 1990 to 6368 (4.7% of all deaths) in 1995, representing an increase of 27% in such cases.[36] If one adds to this number the cases in which treatment was withheld or withdrawn with the explicit intent of shortening life (18,000), the cases in which the lives of

[34] Margaret P. Battin, *The Least Worst Death* (New York: Oxford University Press, 1994), at 132–133. See also Pieter V. Admiral, "Justifiable Euthanasia," *Issues in Law and Medicine*, Vol. 3, No. 4 (Spring 1988): 361–370; H.M. Kuitert, "Euthanasia in the Netherlands: A Practice and Its Justification," lecture delivered at the First World Congress of Bioethics (Amsterdam, 1992).

[35] South Australian Voluntary Euthanasia Society, Fact Sheet 17 (February 1997).

[36] Herbert Hendin, Chris Rutenfrans and Zbigniew Zylicz, "Physician-Assisted Suicide and Euthanasia in the Netherlands," *JAMA*, Vol. 277, No. 21 (June 4, 1997), p. 1721. Van der Maas *et al* object to combining different categories of actions at the end-of-life. Cf. Johannes J.M. van Delden, Loes Pijnenborg and Paul J. van der Maas, "Dances with Data," *Bioethics*, Vol. 7, No. 4 (1993): 323–329, at 325.

neonates were intentionally ended (90 cases), and the cases of psychiatric patients who were assisted in suicide (2 to 5 cases), then the total rises to over 24,000 cases.[37]

The Dutch investigators minimized the significance of the number of deaths without consent by explaining that the patients were incompetent.[38] However, Hendin *et al* argue that in the 1995 study, 21% of the individuals classified as "patients whose lives were ended without explicit request" were competent, whereas in the 1990 study, 37% were reported as competent. They maintain that more than 4000 additional competent patients were given pain medication in amounts likely to end their lives by physicians without discussing this decision with the patients, though the primary intention was not to end their lives. The physicians justified their actions by saying that they had some previous discussion with the patient. Yet, Hendin *et al* conclude, "it seems incomprehensible that a physician would terminate or put at risk the life of a competent patient on the basis of a previous discussion without checking what the patient felt currently."[39]

A study, presented at the 20th International Congress of Chemotherapy in Sydney (July 1997), claimed that the law exempting Dutch physicians from criminal prosecution in cases of strictly controlled euthanasia had created a "slippery slope." It pointed to the hundreds of patients whose lives had been ended without their explicit request through the use of painkillers by physicians in the Netherlands in 1995. One of the study's authors, Dr. Dick Willems, referred to this situation as "very worrisome," while at the same time denying that the situation had worsened since euthanasia's quasi-legalization.[40]

Other commentators were already arguing in the late 1980s that there were grounds for concern about the Dutch experience. Ph. Schepens stated that the performance of euthanasia without the consent of the patient was from two to five times more frequent than euthanasia with consent.[41] Some Dutch studies have indicated that elderly people may be fearful of their lives being ended without their consent and that, in fact, families in Holland request euthanasia more often than the patient.[42]

37 Henk Jochemsen and John Keown, "Voluntary Euthanasia under Control? Further Empirical Evidence from the Netherlands," *Journal of Medical Ethics*, Vol. 25 (1999): 16–21, at 19–20, reprinted in brief in *Issues in Law and Medicine*, Vol. 14 (Spring 1999), at 474. See also J. Keown, "Euthanasia in the Netherlands: Sliding Down the Slippery Slope?," *Notre Dame J. of Law, Ethics and Public Policy*, Vol. 9 (1995): 407–448.

38 Paul J. van der Maas, Gerrit van der Wal, Ilinka Haverkate *et al*, "Euthanasia, Physician-Assisted Suicide, and Other Medical Practices Involving the End of Life in the Netherlands, 1990–1995," *New Eng. J. of Med.*, Vol. 335, No. 22 (November 28, 1996): 1699–1705. See also Loes Pijnenborg, Paul van der Maas, Jan W.P.F. Kardaun *et al*, "Withdrawal or Withholding of Treatment at the End of Life," *Arch. Intern. Med.*, Vol. 155 (February 13, 1995), esp. at 291.

39 Herbert Hendin, Chris Rutenfrans and Zbigniew Zylicz, "Physician-Assisted Suicide and Euthanasia in the Netherlands," *op. cit.*, p. 1721.

40 Steve Dow, "Dutch laws linked to fears over euthanasia," *The Age*: Melbourne Online (July 4, 1997). See http://www.theage.com.au.

41 Ph. Schepens, "Euthanasia: Our Own Future?," *Issues in Law and Medicine*, Vol. 3, No. 4 (Spring 1988): 371–384.

42 J.H. Segers, "Elderly Persons on the Subject of Euthanasia," *Issues in Law & Medicine*, Vol. 3 (1988): 407–424; John Keown, "The Law and Practice of Euthanasia in the Netherlands," *Law Quarterly Review*, Vol. 108 (January 1992): 51–78; R. Fenigsen, "A Case Against Dutch Euthanasia," *Hastings Center Report*, Vol. 19 (Supp.) (1989): 22–30. See also Fenigsen's further critique in "The Report of the Dutch Governmental Committee on Euthanasia," *Issues in Law & Medicine*, Vol. 7, No. 3 (Winter 1991): 339–344; "Physician-Assisted Death in the Netherlands: Impact on Long-Term Care," *Issues in Law & Medicine*, Vol. 11, No. 3 (1995): 283–297, and "Dutch Euthanasia Revisited," *Issues in Law & Medicine*, Vol. 13 (Winter 1997): 301–

Daniel Callahan claims that despite the large number of estimated cases each year, the actual reports that are required by law amount to no more than three hundred. There is no way of knowing whether the other conditions specified by the Dutch Supreme Court are actually being met. There is no effective legal supervision of euthanasia in the Netherlands, and no evidence that the Court's conditions are taken seriously. Moreover, no significant efforts are made by the Dutch police authorities to seek out those practicing euthanasia without reporting it.[43]

The British Medical Association, in its memorandum before the House of Lords regarding the situation in the Netherlands, stated:

> All seem to agree that the so-called rules of careful conduct (official guidelines for euthanasia) are disregarded in some cases. Breaches of rules range from the practice of involuntary euthanasia to failure to consult another practitioner before carrying out euthanasia and to certifying the cause of death as natural. Some would see this as lending credence to the view that even careful circumscription of the practice cannot guarantee observance of the rules.[44]

Paul van der Maas and colleagues reported that the general practitioner consulted a colleague in 81% of cases, whereas in-hospital consultation occurred in 93% of cases.[45] The study does not say whether the consulting colleague was independent from the physician who requested the consultation. Another study holds that consultation takes place in more than sixty percent of euthanasia and PAS cases, in 99% of the cases reported to the public prosecutor and in approximately 37% of unreported cases. In 1990, 7% of GPs met all the criteria for good consultation; this increased to 64% in 1995.[46] No pertinent data is available in the 2001 study. Herbert Hendin argued that consultation in euthanasia cases is often for the sake of the form and that the consulting doctor often does not see the patient at all. In response to criticism about this situation, the Royal Dutch Medical Association has issued revised Guidelines that urge consultants to actually see the patient, stating that a telephone conversation with the referring doctor is insufficient. According to Hendin, consultation offers the patient little protection anyway. In cases where the details are available, the consultant is usually found to be a euthanasia advocate who routinely confirms proceeding with the process without conducting a truly independent evaluation.[47]

311. In his comments on the first draft of this study, Leenen wrote that Schepens, Segers, Fenigsen and Callahan do not base their arguments on research. These are not 'studies.' Letter dated July 25, 2000.

43 Daniel Callahan, *The Troubled Dream of Life* (New York: Simon and Schuster, 1993), at 113. See also Daniel Callahan and Margot White, "The Legalization of Physician-Assisted Suicide: Creating a Regulatory Potemkin Village," *Un. of Richmond L. Rev.*, Vol. 30, No. 1 (January 1996): 1–81, esp. 13–18; Richard Fenigsen, "The Netherlands: New Regulations Concerning Euthanasia," *Issues in Law and Medicine*, Vol. 9, No. 2 (Fall 1993): 167–173; C..I. Dessaur and C. J. C. Rutenfrans, "The Present Day Practice of Euthanasia," *Issues in Law and Medicine*, Vol. 3, No. 4 (Spring 1988): 399–405; Peter A. Singer and Mark Siegler, "Euthanasia – A Critique," *New England J. of Medicine*, Vol. 322 (June 1990): 1881–1883; John Keown, "On Regulating Death," *The Hastings Center Report*, Vol. 22 (March–April 1992): 39–43.

44 House of Lords, *Select Committee on Medical Ethics*, session 1993–94, Vol. II, Minutes of Oral Evidence (London: HMSO, 1994), at 33.

45 P.J. van der Maas, J.J.M. van Delden, and L. Pijnenborg, *Euthanasia and other Medical Decisions Concerning the End of Life*, *op. cit.*, p. 47. See also p. 44.

46 Breje D. Onwuteaka-Philipsen, Gerrit van der Wal, Piet J. Kostense and Paul J. van der Maas, "Consultation with another Physician on Euthanasia and Assisted Suicide in the Netherlands," *Social Science and Medicine*, Vol. 51 (2000): 429–438.

47 Herbert Hendin, "The Slippery Slope: The Dutch Example," *Duquesne L. Rev.*, Vol. 35 (1996), pp. 429–430. See also KNMG Guidelines on Assisted Suicide and Euthanasia (August 1995).

Because of these contrasting interpretations, fieldwork is necessary for the understanding of the Dutch euthanasia policy and practice. Before introducing my research and presenting the interviewees' responses to my questions concerning the worries expressed *supra*, let me provide some further background about the practice of euthanasia and the legal framework.

CHAPTER 2

THE PRACTICE OF EUTHANASIA AND THE LEGAL FRAMEWORK

1. THE PRACTICE OF EUTHANASIA

The three relevant categories of Dutch doctors who are involved in the practice of euthanasia are GPs, nursing-home doctors, and specialists. Every person in the Netherlands has a more or less permanent relationship with a GP, who provides primary health care and is the point of entry for specialist care. GPs have the most extensive experience with euthanasia insofar as they discuss it most frequently with their patients, they receive two-thirds of all requests, and they are generally the most willing to perform it. The level of experience with euthanasia among specialists is about half that of GPs (with 3% of all deaths in their practice attributable to euthanasia). By contrast, euthanasia plays a small role in the practice of nursing-home doctors, who receive relatively few requests (only a fifth of them have ever honored one).[1]

The Guidelines speak of "persistent request." A request made on impulse or as a result of a temporary period of depression should not be honored. The request must have been discussed repeatedly and thoroughly a number of times during several conversations. However, van der Wal and colleagues conducted a survey among a random sample of family doctors, showing that in 22% of cases the request was made only once.[2]

The rate of record keeping[3] and written requests[4] in euthanasia cases improved during the 1990s, but the situation is still unsatisfactory. There are written requests in about 60% and written record keeping in some 85% of all cases of euthanasia.[5] A most troubling phenomenon is the significant number of unreported euthanasia cases. Since November 1990, new state regulations require physicians to report cases of euthanasia to the local coroner and the public prosecutor. The number of reports rose from 454 cases in 1990 to 591 in 1991, to 1323 in 1992, to 1318 in 1993, and to 1424 in 1994. In 1999, the total number of reports was 2216.[6] This considerable increase suggests that more physicians are willing to

1 John Griffiths, "Effective Regulation of Euthanasia and Other Medical Behavior that Shortens Life," in Ejan Mackaay (ed.), *Uncertainty and the Law* (Montreal: Editions Thémis, 1999), pp. 72–73.

2 G. van der Wal, J.Th.M. van Eijk, H.J.J. Leenen and C. Spreeuwenberg, "Euthanasia and Assisted Suicide. II. Do Dutch Family Doctors Act Prudently?," *Family Practice*, Vol. 9, No. 2 (1992), p. 112.

3 A doctor has an obligation to maintain a full dossier on every patient and to accurately record therein what he or she does and why. Keeping adequate records is a general requirement of medical practice, and specifically is one of the requirements of careful practice in the case of euthanasia.

4 Kimsma notes that written requests for euthanasia are preferable but not mandatory. Another acceptable solution is a witness.

5 John Griffiths, "Effective Regulation of Euthanasia and Other Medical Behavior that Shortens Life," *op. cit.*, p. 74.

6 I thank Henk Leenen for this piece of information.

acknowledge and report their actions, having seen that their colleagues are not being prosecuted for performing euthanasia. At the same time, the Remmelink Commission detected 2300 cases of euthanasia, which means that about half are still unreported.[7] John Griffiths writes that the reporting rate for euthanasia[8] was 18% in 1990, and that by 1995 it had risen to 41%. A situation in which less than half of all cases are reported is unacceptable from the standpoint of effective control.[9]

The Dutch approach to euthanasia is said to reflect an open attitude towards tackling a difficult moral issue. For the past twenty five years, the debate has been discussed openly in all circles of society. It has been considered in the Parliament, addressed by the courts, debated in religious institutions, and has required the constant attention of the Royal Dutch Medical Association. It continues to be a focus of the media, and polls have been conducted from time to time to examine public attitudes on this issue.

Despite this apparent openness, the 1990 study shows that 22% of physicians feel that they should not always be required to report euthanasia as unnatural death. The legal ambiguity that existed for twenty years made Dutch doctors feel uncomfortable with reporting euthanasia, citing prosecution as an objection. They emphasized that they would be prepared to report euthanasia as such, but did not wish to be considered as a suspect in a criminal act. Thus, the uncertainty of what might happen to the physician was considered an obstacle to reporting an unnatural death.[10] To address this issue, a careful, clearly stated procedure was needed, one that would be explicitly recognized under the law.

2. THE LEGAL FRAMEWORK

The legal ambiguity was the result of prohibiting euthanasia under the law while allowing the practice under certain circumstances. Two Articles of the Criminal Code explicitly prohibit euthanasia: Article 293 prohibits killing a person at his or her request ("Any person who takes another person's life at that person's express and earnest request shall be liable to a term of imprisonment not exceeding twelve years or a fine of NLG 100,000"); Article 294 prohibits assisted suicide ("Any person who intentionally incites another person to commit suicide, assists him in the act or provides him with the means to commit suicide shall, if suicide follows, be liable to a term of imprisonment not exceeding three years or a fine of NLG 25,000").[11]

Despite these legal provisions, the courts have held that Article 40 of the Criminal Code ("Any person who was compelled by *force majeure* to commit a criminal act shall not be

[7] Henk A.M.J. ten Have, "Euthanasia: The Dutch Experience," *Annals de la Real Academia Nacional de Medicina,* Tomo CXII (Madrid, 1995), pp. 436–437.

[8] Reporting, as opposed to record keeping, refers to the requirement that a doctor report a case of euthanasia to the authorities as an unnatural death.

[9] John Griffiths, "Effective Regulation of Euthanasia and Other Medical Behavior that Shortens Life," *op. cit.,* pp. 74–75.

[10] P.J. van der Maas, J.J.M. van Delden, and L. Pijnenborg, *Euthanasia and other Medical Decisions Concerning the End of Life, op. cit.,* p. 98.

[11] Netherlands Ministry of Foreign Affairs – APPENDICES. For further reading, see H.J.J. Leenen, "Euthanasia, Assistance to Suicide and the Law: Developments in the Netherlands," *Health Policy,* Vol. 8 (1987): 197–206; J.K.M. Gevers, "Legal Developments Concerning Active Euthanasia on Request in the Netherlands," *Bioethics*, Vol. 1, No. 2 (1987): 156–162, and http://www.bz.minbuza.nl/English/Policy/c_eutheng-app.html#

criminally liable")[12] provides a defense to doctors charged under Articles 293 and 294. The *overmacht* defense, which, like *force majeure,* translates as overpowering force, envisions a case of urgency whereby the accused is driven by his or her conscience to commit an offence that amounts to a lesser evil than would have ensued had events been permitted to run their course. As such, the accused decided to make the deliberate moral choice to break the law because the force of circumstances precluded delaying action. However, the defense does not stand if there was a reasonably available option whereby the accused could have avoided the commission of the offence.[13]

A major step was taken in 1990 on behalf of physicians practicing euthanasia. In soliciting for the approval and cooperation of the Royal Dutch Medical Association with the Remmelink study, the Ministry of Justice not only promised legal immunity for physicians participating in the national investigation, but it also agreed to proclaim a notification procedure that included the following elements:

(a) The physician performing euthanasia does not issue a declaration of a natural death and informs the local medical examiner by means of an extensive questionnaire;
(b) The medical examiner reports to the district attorney;
(c) The public prosecutor decides whether a prosecution must be started. As a general rule, cases in which the doctor has complied with the requirements for euthanasia would not be prosecuted.[14]

The notification procedure was granted a formal legal status by a procedural law that came into force on June 1, 1994. According to this law, a physician performing euthanasia in compliance with the criteria that have been developed in case law and medical ethics will, as a general rule, not be prosecuted.[15]

On November 28, 2000, the Dutch Lower House of parliament, by a vote of 104 for and 40 against, approved the legalization of euthanasia. On April 10, 2001 the Dutch Upper House of parliament voted to legalize euthanasia, making the Netherlands the first and at that time only country in the world to legalize euthanasia. Forty-six members of the 75-seat Senate voted for the *Termination of Life on Request and Assistance with Suicide Act*; twenty-eight voted against; one member was not present. A year later, in April 2002, the legalization process was completed when the law was approved by the Dutch Senate.

The new legislation makes it legal to end a patient's life, subject to the following criteria: the patient must be suffering unbearable and unremitting pain, with no prospect of improvement. The patient must make a sustained, informed and voluntary request for help to

12 http://www.bz.minbuza.nl/English/Policy/c_eutheng-app.html#

13 Barney Sneiderman and Marja Verhoef, "Patient Autonomy and the Defence of Medical Necessity: Five Dutch Euthanasia Cases," *Alberta Law Review,* Vol. XXXIV, No. 2 (1996), p. 376. In England and Canada, the defense of medical necessity has been recognized in abortion cases (see *R. v. Bourne* [1938] 3 All E.R. 615 and *R. V. Morgentaler,* [1975] 20 C.C.C. 2d, 449), but it is not available in euthanasia cases. See *R. v. Cox* [1992], unreported. R. Porter, "Doctor convicted of attempted murder," *Sunday Telegraph* (September 20, 1992), and the Latimer case, http://www.cnn.com (November 5, 1997); *Reuters*, Toronto, "Canada farmer gets two years for mercy killing of daughter" (December 1, 1997).

14 Robert J.M. Dillmann and Johan Legemaate, "Euthanasia in the Netherlands: The State of the Legal Debate," *European J. of Health Law,* Vol. 1 (1994), p. 84.

15 A translation of the law can be found in John Griffiths, Alex Bood and Heleen Weyers, *Euthanasia and Law in the Netherlands, op. cit.*, pp. 308–313.

die. All other medical options must have been previously exhausted. A second medical opinion must be sought to confirm diagnosis and prognosis. The termination of life must then be carried out in a medically appropriate care and attention. The physician is obliged to report the death to the municipal pathologist, specifying whether the cause of death was euthanasia or assisted suicide.[16]

Doctors will be immune from prosecution for helping a patient to die, as long as they follow this set of Guidelines. They will still report cases of voluntary euthanasia to the coroner and a regional committee, who can recommend prosecution leading to a prison sentence of up to 12 years if the Guidelines have not been followed. The new Act changed the emphasis on who should prove guilt or innocence if the code of practice is breached. Previously, the onus was squarely on the doctors to prove that they had followed the Guidelines and were therefore innocent of any offence. However, the new law shifts the responsibility for proving guilt to the regional committees.

The law contains special provisions dealing with requests from minors for termination of life and assisted suicide. The most controversial aspect of the original act was that incurably ill minors between the ages of 12 and 16 may request and receive help to die, with the agreement of their parents. In exceptional circumstances, doctors may even be able to help the child to die without parental consent, although such cases are likely to be rare. Persons of 16 to 18 years of age would be able to request euthanasia without recourse to their parents' approval.[17]

In July 2000, in response to critical questions by members of parliament, the Cabinet dropped the provision that euthanasia requests by minors between 12 and 16 years in exceptional cases could be granted without the parents' consent. Some analysts viewed this retreat as a manoeuvre to win approval for other controversial provisions of the new legislation, such as legalizing euthanasia for victims of Alzheimer's disease.[18] Still, allowing euthanasia for minors 12 years of age and older seriously overestimates the capacity of minors to evaluate the meaning and consequences of a request to die. It places an unacceptable burden on these young people and may well disturb society's confidence in the relationship between physicians, parents and children. Jochemsen rightly says that unless we are prepared to give minors the right to do everything else in life that an adult can do, giving them the right to end life seems out of place.[19]

The new law also establishes a legal basis for advance euthanasia declarations via a type of living will in which competent patients would request euthanasia in the event they become mentally incompetent. Though such a statement does not imply that a physician has a duty to perform euthanasia, it provides the legal opening to end the life of incompetent patients who had signed such a document.

The Royal Dutch Medical Association (KNMG) welcomed the legislation process. It has long argued for ending the paradoxical legal situation that physicians acting within the

[16] *Reuters*, "Dutch Parliament votes to legalize euthanasia" (November 28, 2000).

[17] "Minderjarige mag euthanasie vragen," *NRC Handelsblad* (July 10, 1999), p. 3. See also Marilyn Gardner, "Dutch poised to legalize euthanasia," *The Christian Science Monitor* (June 30, 2000). http://www.csmonitor.com/durable/2000/06/28/text/p1s4.html

[18] "Dutch call off aided suicide for children," *The International Herald-Tribune* and *The Associated Press* (July 15, 2000).

[19] Henk Jochemsen, "Update: The Legalization of Euthanasia in The Netherlands," *Ethics & Medicine*, Vol. 17, No. 1 (2001): 7–12.

criteria for careful conduct could still face criminal prosecution. The Justice Minister, Benk Korthals, emphasized that physicians must be convinced of the "objective medical reasons" for the patient's suffering. Being merely "tired of life" was not sufficient.[20]

2.1. Leading Court Cases

The first case in which a physician was found guilty under Article 293 but not punished took place in the early 1950s. It concerned a patient who had been suffering from advanced tuberculosis and had for some time been urging his brother, who was a physician, to end his misery. Finally, the brother administered painkillers and sleeping pills and killed the patient. The physician was prosecuted, convicted and received a one-year probation. Apparently, the court considered the relationship between the physician and his brother, as well as the patient's terminal illness, as mitigating factors.[21]

In 1966 a young woman named Mia Versluis suffered cardiac arrest under anesthesia and became comatose. After five months, the anesthesiologist, who had continued to care for her, went to remove her from the ventilator. Her father, who believed mistakes had been made during the original anesthetic, lodged a complaint with the Medical Disciplinary Tribunal (equivalent to our Board of Medical Examiners), and the case went through the court system. The court ruled that in termination of life support, other colleagues must be consulted first, and the situation must be discussed with the family. The doctor was fined 1000 guilders and the court ruling was made public (relatively severe sanctions under Dutch disciplinary law).[22] However, it was not until the *Postma* case that the euthanasia movement was really put in motion.

2.1.1. Postma

The campaign to legalize physician-assisted suicide in the Netherlands began in 1973 with this case, in which a family doctor was prosecuted for giving a lethal injection of morphine to her mother, who lived in a nursing home. Ms. Postma, a 78-year-old widow, had been in a nursing home since a cerebral hemorrhage had left her paralyzed on one side. On several occasions, she had asked her daughter to end her life. She had also spoken of not wanting to live any more to her other daughter and to the nursing home staff. Finally, after many repeated requests, her daughter acquiesced in the presence of the daughter's husband, who was also a physician. She was promptly charged with manslaughter, which stirred up a great deal of public sympathy for the daughter. After two years, the court found her guilty under existing law, ruling that the lethal injection was not a reasonable means to put an end to her mother's suffering. At the same time, the court acknowledged that doctors are not obliged to prolong life at any cost and that under certain conditions it can be legitimate to use medication with the intention of shortening life. Consequently, the verdict was very lenient, with Dr. Postma given a

20 Tony Sheldon, "Netherlands Gives More Protection to Doctors in Euthanasia Cases," *British Medical Journal*, Vol. 321 (December 9, 2000), p. 1433.

21 Jim Persels, "Forcing the Issue of Physician-assisted Suicide," *Journal of Legal Medicine*, Vol. 14 (1993), p. 105.

22 Joseph B. Vander Veer Jr., "Euthanasia in the Netherlands," *Journal of the American College of Surgeons*, Vol. 188, Issue 5 (May 1999): 532–537.

conditional jail sentence of one week with a one-year probation. Later, even that sentence was suspended.[23]

Undoubtedly, this was a highly emotional case. The procedure, however, was very problematic. From the details of the case it seems that no independent physician was consulted and that all decisions remained in the family. The extent to which close members of the family are able to make unbiased, non-partisan and professional decisions is questionable. The control mechanisms of the euthanasia practice require including uninvolved professionals.

Following the *Postma* decision, the KNMG (Royal Dutch Medical Association) issued a statement that the administration of pain relieving drugs and the withholding or withdrawal of futile treatment could be justified even if it resulted in death. In 1981, a lay volunteer, named Ms. Wertheim, assisted in the suicide of a non-terminally ill patient, who on many occasions had expressed her wish to die. The patient's GP had refused to accede to her request and had referred her to Ms. Wertheim, who, after a few meetings, assisted with the patient's suicide. She was prosecuted in the Rotterdam Criminal Court, which acknowledged the patient's right to self-determination and set forth standards for non-criminal aid in dying. The most noteworthy of these were that a patient requesting the aid need not be terminally ill, but need only be experiencing unremitting physical and mental suffering, and that a doctor must be involved in the decision and must determine the method to be used. As a jail sentence would have been too much of a burden for the 76-year-old Ms. Wertheim to bear, she was given a conditional sentence of six months subject to a one-year probation.[24]

2.1.2. Schoonheim

The first acquittal of a doctor committing euthanasia took place in 1983 and was upheld by the Supreme Court (called Hoge Raad) in the *Schoonheim* case in 1984. The case concerned Maria Barendregt, a 95-year-old bedridden patient who was totally dependent on the nursing staff for her bodily needs and who continuously expressed her desire to die. Her doctor consulted with another physician, who concurred that the patient was unlikely to regain her health. However, although the patient was suffering, her illness was not terminal. The Alkmaar Court diverged from the existing criteria of *continuous and unbearable* suffering of a *physical and spiritual* nature, and instead required the defense to show that the patient was only under *continuous* suffering and that euthanasia could be justified solely on the basis of *psychological* suffering.[25] The prosecution appealed to the Court of Appeals in Amsterdam, which, in turn, rejected Schoonheim's defense and found him guilty, holding that his conduct constituted an offence under Article 293 of the Penal Code. At the same time, the Court used its discretion not to impose any punishment, given the judges' assessment that the doctor

[23] Charles F. McKhann, *A Time to Die: The Place for Physician Assistance* (New Haven, Conn.: Yale University Press, 1999), p. 122. The *Postma* case was the best known prosecution during this period of a person who killed another at the latter's request, but it wasn't the only one. There were at least three other prosecutions for violations of Articles 293 or 294. Cf. Carlos F. Gomez, *Regulating Death* (New York: The Free Press, 1991), pp. 28–32; John Griffiths, Alex Bood and Heleen Weyers, *Euthanasia and Law in the Netherlands, op. cit.*, p. 53.

[24] Jim Persels, "Forcing the Issue of Physician-assisted Suicide," *op. cit.*, p. 106; John Griffiths, Alex Bood and Heleen Weyers, *Euthanasia and Law in the Netherlands, op. cit.*, pp. 58–59.

[25] Jim Persels, "Forcing the Issue of Physician-assisted Suicide," *op. cit.*, p. 107. See also Joseph B. Vander Veer Jr., "Euthanasia in the Netherlands," *op. cit.*

acted with integrity and due caution. On further appeal, the Supreme Court affirmed the appellate court's absence of material illegality analysis. However, it remanded the case to the Court of The Hague to determine if Schoonheim's conduct was justified under the defense of necessity (Article 40). The Court accepted the doctor's defense that he faced a conflict of responsibilities between preserving the patient's life and alleviating suffering. In cases where the doctor is confronted by such a conflict and the objective circumstances justify shortening the patient's life, the resolution of the conflict might necessitate the performance of euthanasia. This decision would have to be reached on the basis of the doctor's responsible medical opinion, as measured by the prevailing standards of medical ethics.[26] The Hague Court of Appeals acquitted Schoonheim, deeming that he had properly resolved his conflict of duties.

The reading of the details of this case suggest that there were compelling reasons to perform euthanasia. First, the patient's decision was voluntary. Mrs. Barendregt agonized over her progressive degradation as she became steadily less able to communicate with her loved ones and the medical staff. By the last week of her life, she was no longer able to take liquids or to speak, and she was suffering from periodic lapses of consciousness. However, when she regained some ability to communicate, she adamantly expressed her horror at the fact that she was still alive. Available medical treatment did not seem to help her. Mrs. Barendregt was fully aware of her condition, which brought her to pray for death. There was no indication that her family exerted pressure on her to die. She was the one who begged her son to urge the physician to end her life at the earliest possible moment. The patient had a long farewell with her son and daughter-in-law, at which time she expressed her gratitude to Dr. Schoonheim for his devoted care and for agreeing to terminate her life. Finally, the case was documented, as it should be.

Having said that, the decision-making process remains somewhat questionable. Although it involved Mrs. Barendregt, her son, and the two physicians, more should be done to verify the patient's diagnosis and to see that euthanasia is the *last resort,* after exhausting all treatment alternatives. If the series of talks involved only Dr. Schoonheim and his assistant, a young physician working in his office, then it contravenes the requirement of independent consultation. In order to minimize misdiagnosis and to allow the discovery of other medical options, the decision-making process should include a second opinion provided by a specialist who is not dependent on the first doctor, either professionally or otherwise. Furthermore, it is preferable to broaden the decision-making team to include a lawyer, who can examine the legal aspects of the case; a social worker, who can assess the relationships within the family and verify that the euthanasia request is voluntary and free of coercion; and a psychologist, who can evaluate the patient's frame of mind. Possibly a public representative should be included as well. This extra caution should ensure that the right to die with dignity does not become a duty.

[26] *Nederlandse Jurisprudentie,* 1985, no. 106. John Griffiths, Alex Bood and Heleen Weyers, *Euthanasia and Law in the Netherlands, op. cit.,* pp. 18–19, 62–63, 322–328; B. Sneiderman and M. Verhoef, "Patient Autonomy and the Defence of Medical Necessity: Five Dutch Euthanasia Cases," *op. cit.*, pp. 388–392; Julia Belian, "Deference to Doctors in Dutch Euthanasia Law," *Emory Int. L. Rev.*, Vol. 10 (Spring 1996): 255–295, in URL: http://www.law.emory.edu/EILR/volumes/spring96

2.1.3. Admiraal

The decision in *Schoonheim* was followed by the *Pols* case,[27] and these two court rulings led to a series of judicial decisions through which the conditions and limitations of the defense were gradually worked out. In 1985, the District Court of The Hague acquitted Dr. Pieter Admiraal for the charge of offence under Article 293. The case concerned Karin L., a 34-year-old patient who had been afflicted with multiple sclerosis since 1976. The progressive deterioration of her condition led to a nursing home admission in 1981. By mid-1983, she was completely paralyzed except for the movement of her left hand. Karin was dismayed by her state of total dependency, remarking that she could not even brush away a fly. Despite the fact that Karin's condition was not defined as "terminal," she could barely swallow or speak above a whisper, her sight was failing, and she was experiencing intractable bone pain. Her breathing was becoming more labored, and although she feared suffocation, Karin was appalled at the thought of artificial ventilation. She repeatedly asked to die, knowing that her intolerable condition bore no prospects for improvement. She asked her nursing home physician to end her life, and when he refused she approached the Dutch Voluntary Euthanasia Society for help. Then Dr. Admiraal entered the scene.

The decision-making process was extensive but not flawless. Dr. Admiraal consulted with the terminal care team at his hospital, which cannot be deemed as independent. There were a number of sessions in which he and the team conferred with Karin and her parents. At Dr. Admiraal's request, her family physician of thirty years had a long talk with Karin about her request for euthanasia. The family pastor was also consulted and informed Karin that he did not oppose her decision.[28] On November 4, 1983, Dr. Admiraal performed euthanasia with Karin's family at her bedside. He then reported the case to the police as an unnatural death.

The public prosecutor responded by charging him with failure to consult an expert neurologist on multiple sclerosis prior to acting. The District Court disagreed, holding that Dr. Admiraal had scrupulously complied with the euthanasia Guidelines stipulated by the Supreme Court in the *Schoonheim* case. The Court ruled that when confronted by a situation of necessity, he had carefully weighed the conflicting duties and made a justifiable choice.[29] All of the factors supporting Dr. Admiraal's actions convinced the Court to discount the importance of independent consultation with a neurologist and to acquit him.

Although Karin had thought about ending her life for some time, it was said that she initially decided against taking extreme measures only because she did not want to traumatize her parents.[30] But in the end, the decision to die was undoubtedly made voluntarily. Clearly, her pain was not only physical but mental as well. After seven years of struggling with the disease, Karin was simply exhausted and did not want to continue living such a dependent life.

[27] The case is described in detail by Griffiths *et al.*, *Euthanasia and Law in the Netherlands, op. cit.*, pp. 63–65.

[28] *Ibid.*

[29] *Nederlandse Jurisprudentie* 1985, no. 709; Griffiths *et al.*, *Euthanasia and Law in the Netherlands,* pp. 66–67.

[30] B. Sneiderman and M. Verhoef, "Patient Autonomy and the Defence of Medical Necessity: Five Dutch Euthanasia Cases," *op. cit.*, p. 392.

2.1.4. Duintjer

That same year, on October 4, 1985, 50-year-old Martha N. died after ingesting a lethal dose of cyclobarbital provided by her psychiatrist, Dr. Duintjer. Martha had a history of depression and alcohol abuse dating back about 25 years. She was diagnosed as suffering from a character neurosis, with depressive and dependent traits and a strong self-deprecation leading to suicidal ideation. During 1983–1984, she tried to commit suicide three times, unsuccessfully. She decided to isolate herself in her bedroom, often denying access to her husband and children.

Martha repeatedly asked Dr. Duintjer and her family physician, Dr. W., to assist in her suicide, describing her own life as "one big black hole." Dr. W. arranged for Martha's pastor to talk to her, and after about a dozen meetings he reported that he had come to believe that her case was hopeless and that the physicians were morally entitled to assist in her death. On October 3, 1985 she signed the following statement: "I declare that life has no more value to me and that I voluntarily and in full consciousness, at my time and with the method of my choice, wish to end this life." When Dr. Duintjer reported the assisted suicide, the public prosecutor charged him with aiding suicide under Article 294.[31]

The Court found that the medico-legal criteria had been satisfied: The patient was mentally competent and was acting out of her free will; she had made persistent requests to die; and there were no alternative measures that could relieve her intolerable suffering. The prosecution appealed the decision, but the acquittal was upheld by the Hague Court of Appeals. The Appellate Court rejected the prosecution's argument that psychiatric patients were by definition mentally incompetent to qualify for assisted suicide. The Court held that Martha was mentally competent and that there were no treatment prospects likely to better her situation. The Court admonished the accused for failing to consult a physician not involved in the patient's treatment. However, it let the verdict stand because of the compelling testimony by psychiatrists and psychologists, who had been treating her for years, that Martha's suffering was indeed unbearable and irremediable.[32]

It was said that her husband and children were caring and supportive over the years but that Martha was still unhappy and preferred not to have their company. It was also said that she desperately wanted to die, not only for her own sake but also to put an end to the constant stress and turmoil that her condition imposed upon her family.[33] This sort of consideration deserves special attention to ensure that the patient's decision is not a result of familial or environmental pressures. The description of the case does not mention the option of moving Martha out of her house to a new location. Perhaps there were other available options to rescue Martha from her depression. The extent to which we can speak of "free will" in a deeply depressed person is an open question.

Moreover, the decision-making process was insufficient. The two doctors that were heavily involved in her treatment did not consult another physician, who could have brought a fresh outlook and new insight to the situation. It was appropriate to involve the pastor, but it is still unclear why another professional psychiatrist was not consulted. It is preferable to include members of the relevant professions (social work, psychology, and law), as well as a public representative.

31 *Ibid.*, pp. 396–397.
32 *Ibid.*, at 398.
33 *Ibid.*, p. 397.

2.1.5. Dr. K

Mrs. M., a 73-year-old patient with multiple sclerosis, started talking about euthanasia upon suffering a major deterioration in her health. She had lived a difficult life, but always had continued fighting to gain control over her life. In 1982, fighting seemed to be totally useless. Unable to change the course of her disease, she refused any of the alternative medications proposed by Dr. K., her psychiatrist and friend. On August 4, 1982, Dr. K. committed euthanasia on Mrs. M. and informed her action to the prosecutor.[34]

On March 1, 1984, the District Court of Groningen argued that it is possible to find some criminal act not liable to punishment if it is a medical action, if it is necessary for medical reasons or of critical importance for adequate medical care, and if it is required by prudent medical science. Whether this is the case depends on five preconditions. Accordingly, euthanasia may be performed:

(1) by a physician after consultation with another physician, who has personally seen the patient;
(2) on a patient whose condition is irreversible and who is suffering unbearably;
(3) when the explicit and earnest request can be considered lasting and based on a proper evaluation by the patient of his/her own condition and the available alternatives;
(4) when the patient does not think that there is a reasonable alternative;
(5) when all other requirements of prudence are fulfilled.[35]

In this case, the first precondition was not fulfilled. Dr. K. had merely informed other physicians and discussed the case of Mrs. M. with them. However, none of these physicians had actually seen the patient. Further, the psychiatrist evidently knew her patient well, but at the same time she was a personal friend of Mrs. M. This friendship might have overshadowed her judgment and ability to make a decision free of biases. As in *Postma,* the control mechanisms of the euthanasia practice should include independent, uninvolved professionals. Dr. K. was found guilty and liable to punishment. Upon appeal to the High Court of Leeuwarden, the Court affirmed the conviction under Article 293 of the Criminal Code. Dr. K. further appealed to the Supreme Court, and the conviction was affirmed yet again.[36]

2.1.6. Kors

On October 31, 1990, Maria S. was assisted to die by her longtime pediatrician, Dr. Kors. Maria was a 25-year-old anorexic, weighing no more than 19 kilos. She persisted in her requests for assisted suicide. Determined to die, Maria had resolved to refuse further tube feeding and hospitalization. She was vomiting incessantly and experiencing severe stomach pains. In light of her condition, Dr. Kors decided to help her. After 16 years of treatment, he saw no likelihood that her condition would improve. Dr. Kors consulted a psychiatrist, who had seen Maria on numerous occasions and reported that Maria was mentally competent and that there were no treatment options that held the prospect of easing her suffering.

[34] Jos V.M. Welie, "The Medical Exception: Physicians, Euthanasia and the Dutch Criminal Law," *Journal of Medicine and Philosophy,* Vol. 17 (1992), p. 430.

[35] *Ibid.,* at 431.

[36] *Ibid.,* pp. 431–432.

Maria's decision to opt for death was influenced by her familial circumstances. At about the time that she managed to achieve something in her life – graduating from high school and finding work as a lab technician – her parents divorced. Then her younger brother Ernst became seriously depressed, and some years later he committed suicide. At his funeral, six months before her own death, Maria announced her intention to follow suit. She told Dr. Kors: "When Ernst died, I knew that finally I was allowed to go as well. I cannot go on any longer. I want salvation now. Therefore, I am asking you as a friend to help me. Please help me."[37]

When Dr. Kors finally agreed to provide the drugs to enable her to commit suicide, Maria signed her last will and then arranged for a videotape in which she explained the reasons for ending her life. Dr. Kors reported her death to the authorities, and the public prosecutor responded by charging him with aiding her suicide under Article 294.[38]

The Court was satisfied with the evidence that the medico-legal Guidelines had been fulfilled: The patient was free of psychosis and was otherwise mentally competent; her wish to die was well considered, persistent and free; her suffering was relentless and intolerable; and there were no reasonable treatment regimens to ameliorate her condition. The Court accordingly granted the motion to dismiss the appeal of the public prosecutor.[39]

The last case to be examined also resulted from exceptional familial circumstances that led the patient to request assistance in dying.[40] As it is a highly controversial case, I will review it in more detail. Like all of the previous cases, this case also involved a woman.[41] Unlike the other cases, however, it was the first time that the Supreme Court considered assisted suicide for a psychiatric patient. It should be noted that physician-assisted suicide in psychiatric practice has been estimated to occur two to five times a year in the Netherlands.[42]

2.1.7. Chabot

Unlike Martha N. and Maria S., Hilly Bosscher was a 50-year-old patient with no history of psychiatric disorder. She wanted to die because she felt that her life had lost its meaning after

37 B. Sneiderman and M. Verhoef, "Patient Autonomy and the Defence of Medical Necessity: Five Dutch Euthanasia Cases," *op. cit.*, pp. 393–394.

38 *Ibid.*, pp. 393–395.

39 *Ibid.*, p. 396.

40 Two important precedents, *Prins* and *Kadijk*, not considered here were concerned with the termination of life of severely defective newborn babies. Cf. Griffiths *et al.*, *Euthanasia and Law in the Netherlands, op. cit.*, pp. 83–84, 341–351. For further deliberation, see A. van der Heide, P.J. van der Maas, G. van der Wal *et al*, "Medical End-of-life Decisions Made for Neonates and Infants in the Netherlands," *Lancet*, Vol. 350 (1997): 251–255; Henk Jochemsen, "Dutch Court Decisions on Nonvoluntary Euthanasia Critically Reviewed," *Issues in Law & Medicine*, Vol. 13, No. 4 (1998), esp. pp. 450–458; Arlene Judith Klotzko, "What Kind of Life? What Kind of Death? An Interview with Dr. Henk Prins," in David C. Thomasma, Thomasine Kimbrough-Kushner, Gerrit K. Kimsma, and Chris Ciesielski-Carlucci (eds.), *Asking to Die, op. cit.*, pp. 388–406; Joseph B. Vander Veer Jr., "Euthanasia in the Netherlands," *op. cit.*

41 Dr. Herbert Cohen, a well-known practitioner of euthanasia in the Netherlands, explains that all of the cases that have broken new ground in Dutch law involved women by saying that women can make an appeal to a doctor that is stronger, more existential. See Herbert Hendin, "Seduced by Death: Doctors, Patients and the Dutch Cure," *Issues in Law and Medicine*, Vol. 10, No. 2 (1994), p. 137. In general, more euthanasia cases are found in women than in men. Cf. Gerrit van der Wal and Paul J. van der Maas, "Empirical Research on Euthanasia and Other Medical End-of-Life Decisions and the Euthanasia Notification Procedure," *op. cit.*, p. 157.

42 Johanna H. Groenewoud, Paul J. van der Maas, Gerrit van der Wal *et al*, "Physician-assisted Death in Psychiatric Practice in the Netherlands," *New Eng. J. of Medicine*, Vol. 336, No. 25 (June 19, 1997), p. 1797.

the death of her two sons: Peter from suicide at the age of 20, and Robbie five years later from cancer, also at the age of 20. In May 1991, on the day of Robbie's death, Mrs. Bosscher tried to commit suicide, unsuccessfully. After her personal family doctor, as well as some other people she knew, refused to help her commit suicide, she approached the Dutch Voluntary Euthanasia Society for help, and they referred her to a psychiatrist, Dr. Chabot.[43] Dr. Chabot conducted a "trial therapy" with Mrs. Bosscher, consisting of a series of thirty sessions of 55 minutes each over a two-month period. However, Mrs. Bosscher told Dr. Chabot that she was not prepared to undertake the commitment to work with him to change her bleak outlook on life. In her personal diary she wrote:

> I have lost everything and will never get it back. I do not want to become another person than I was when I was a mother and happy. It is finished, it is all over. For me alone there is no purpose in life. I know who I am or what I am. To become so different that I will want to or have to live means to me that I have to lose again. I am not allowed to be who I am or was. That's not right.[44]

In a letter to Dr. Chabot in September 1991, after he told her that he would assist in her suicide, Mrs. Bosscher wrote:

> I feel so 'happy' with the help in dying I'll receive. I got everything in my life with which I couldn't possibly live any further. Am I egoist to not want, nor not to be able, to live on? Not having the urge or power to endure, to fight?... I am certainly making life difficult for you. You told me that was none of my business. You wanted to 'invest' in me (I don't know how to express this in a better way). You are a psychiatrist, and as such you tried everything to hold me. But I feel you tried as a human being as well. I must have been a heavy load to take on.[45]

I asked Dr. Chabot what Mrs. Bosscher meant by her use of the term "invest." In his personal letter, Chabot's response was that Mrs. Bosscher felt that he really wanted her to go through her vale of grief. He offered her intensive grief therapy, not just in an impersonal way but also in a way that showed he cared about her as a human being. Mrs. Bosscher "was not just an egocentric who could not perceive" Chabot's intentions; "she had felt that, as a professional, I wanted 'to invest' (time, energy, sharing her pain, etc.) in her." This she could not or would not accept.[46]

The short but intense acquaintance with Mrs. Bosscher (from August 3 to September 7, 1991) led Dr. Chabot to conclude that she was a mentally competent person whose freedom of choice was not constricted by mental illness. In his opinion, Mrs. Bosscher had been suffering from a complicated grief process for five years following the suicide of her son, Peter, in 1986. Chabot did not see any psychiatric illness, clinical depression, trace of psychosis, or personality disorder. He believed that for her, there was no future without her children. He accepted that suicide was the only option to end Mrs. Bosscher's misery and was convinced that she would kill herself in any event, with or without his help. Chabot tried to give her antidepressant medication but she refused, saying that "the only sense life has got for me now is to find my way to Peter and Robbie through a dignified death."[47]

[43] Herbert Hendin, *Seduced by Death* (New York: W.W. Norton, 1997), pp. 60–61.

[44] B. Sneiderman and M. Verhoef, "Patient Autonomy and the Defence of Medical Necessity: Five Dutch Euthanasia Cases," *op. cit.*, p. 400. For further account, see Herbert Hendin, "Seduced by Death: Doctors, Patients and the Dutch Cure," *op. cit.*, pp. 145–152.

[45] I am grateful for Dr. Chabot for sending me (on July 16, 2000) excerpts from Mrs. Bosscher's farewell letter.

[46] Chabot's letter dated August 14, 2000.

[47] "Arlene Judith Klotzko and Dr. Boudewijn Chabot Discuss Assisted Suicide in the Absence of Somatic Illness," *Cambridge Quarterly of Healthcare Ethics*, Vol. 4 (1995), pp. 241–242.

Strangely, in his discussion with Arlene Klotzko, Chabot said that he did insist on a trial with antidepressants before he agreed to assist in her suicide.[48] Yet, later in the same discussion, when asked whether Mrs. Bosscher rejected the option of antidepressant medication, Chabot answered: "This question I consider to be very tendentious given her grief. I hope that no pill will be discovered that will prevent or cure grief. Certainly, antidepressants don't cure grief."[49] I asked Dr. Chabot about this puzzle, and his answer was that he did insist on antidepressants for the few symptoms of depression that he noticed. He felt that the patient should give them a serious try. But, Chabot explained, depression is not identical with grief. It was Mrs. Bosscher's grief that he considered to be by far the most important in her wish to die. Chabot testified that he has had quite a lot of experience in grief therapy and has been successful in that field, "but never with pills." Hence, urging Mrs. Bosscher to try antidepressants for what Chabot conceived as "mild depression" seemed to him perfectly compatible with his hope that there would be no medicine discovered that *by itself* (Chabot's emphasis) would cure grief.[50]

Dr. Chabot had transcribed all the sessions with Mrs. Bosscher, which he sent to four psychiatrists and a clinical psychiatrist. He also consulted a family physician and a theologian-ethicist. He then held lengthy telephone conversations with the consultants, four of whom he met with in person. He asked one of the psychiatrists to meet with Mrs. Bosscher in person, but the colleague declined because Dr. Chabot's extensive documentation of the case had convinced him that it was not necessary. All save one reported that it was unlikely that anything could be done to make Mrs. Bosscher's life bearable and that they would support his decision to assist in her suicide. The psychiatrist who expressed a contrasting view thought that Mrs. Bosscher's condition was not hopeless and that Dr. Chabot should persist in treating her.[51]

In a personal communication, Dr. Chabot wrote that the account of the case, as described by Barney Sneiderman and Marja Verhoef, is accurate.[52] The immediate questions that come to mind are: Why did the four experts, who read the detailed transcripts, take the time to meet with Dr. Chabot but not see a need to meet with Mrs. Bosscher? In other words, if the transcripts were so straightforward, to the extent of making a meeting with the patient redundant, why was there a need to meet with the doctor? Furthermore, was the lone dissenter asked to meet with Mrs. Bosscher? What was his reaction? Did he refuse as well? Maybe he could have saved her life.

Hendin argues that Chabot asked only Dr. Frank van Ree, one of the few Dutch psychiatrists publishing on assisted suicide, to see Mrs. Bosscher. Van Ree felt that this was

48 *Ibid.*, p. 244.
49 *Ibid.*, at 246.
50 Chabot's letter dated August 14, 2000.
51 B. Sneiderman and M. Verhoef, "Patient Autonomy and the Defence of Medical Necessity: Five Dutch Euthanasia Cases," *op. cit.*, p. 402. According to Hendin, *two* of the experts did not recommend Dr. Chabot to assist in her suicide. See "Seduced by Death: Doctors, Patients and the Dutch Cure," *op. cit.*, p. 147. I asked Chabot about this discrepancy, and he explained "both Sneiderman *et al.* and Hendin are right; they simply refer to different moments in the process." Sneiderman *et al.* refer to the period when Mrs. B. was still alive, whereas Hendin refers to the later phase, when the Medical Disciplinary Board invited another expert to give his opinion and he disagreed with Chabot. Strictly speaking, this latter expert was not consulted by Chabot. Letter dated August 14, 2000.
52 Personal communication by Dr. Chabot, dated June 5, 1999.

unnecessary.[53] If this information is correct, it is like inviting the "right" answer rather than seeking professional evaluation of the patient's condition. Later, the court asked van Ree why he did not see the patient. Van Ree explained that he felt he knew the case and that it would only cause the patient further pain to be seen again by someone else. Presumably, Hendin writes, it was less harmful to Mrs. Bosscher to help her commit suicide.[54]

Given the near unanimity of opinion, Dr. Chabot felt assured that he could in good conscience assist Mrs. Bosscher's suicide. Sneiderman and Verhoef wrote that still Chabot sought further counsel from Dr. V., a family physician whose clinical judgment he highly respected. After lengthy discussions on the case, Dr. V. agreed that Mrs. Bosscher's frame of mind precluded any change of heart. However, he was not asked to examine the patient.[55]

This is striking and strange. Why not? It seems that the purpose of the meeting was to calm the conscience of Dr. Chabot and to reassure his decision, rather than to seek an independent and free opinion. It seems that Dr. Chabot was seeking not just any opinion, but a certain opinion, namely, one that conformed to the decision that he apparently had already made to help Mrs. Bosscher end her life.

On September 28, 1991, Dr. Chabot assisted Hilly Bosscher to commit suicide and subsequently stood trial. This was a little over two months after their first meeting and about four months after the death of her younger son. The Assen District Court acquitted him in April 1993 after becoming convinced that Mrs. Bosscher was experiencing long-term psychic suffering that for her was unbearable and unremitting.

Four of the clinical experts consulted by Dr. Chabot appeared as witnesses for the defense. They all testified that the case was so well documented that it was "highly unlikely" that any new information would have been gained by interviewing the patient. The Court then consulted three additional experts who provided written testimonials. None of the seven experts expressed disagreement with Dr. Chabot's role in Mrs. Bosscher's suicide,[56] a fact which critics of the Dutch policy on euthanasia may take to be worrisome and disturbing.

The Appeal Court of Leeuwarden upheld Dr. Chabot's acquittal, but on June 21, 1994, the Supreme Court reversed the decision and convicted Dr. Chabot under Article 294 of the Penal Code. The Supreme Court accepted the contention of the public prosecutor that the defense of *overmacht* was not allowable because none of the experts consulted by Dr. Chabot had personally examined the patient. The Supreme Court held that in non-somatic cases (*i.e.*, cases that have to do with the psychological rather than the physiological condition of the patient), the absence of a face-to-face examination leads to the conclusion that, as a matter of law, the physician may not have truly acted in a state of necessity. Thus, personal contact between consultant and patient constituted a pre-condition to the acceptance of the *overmacht* defense in such a case.

Although Dr. Chabot was found guilty under Article 294 of the Penal Code, the Supreme Court exercised its prerogative under Article 9(a) of the Criminal Code not to impose any

53 Hendin, "Seduced by Death: Doctors, Patients and the Dutch Cure," *op. cit.*, p. 147. See also "Arlene Judith Klotzko and Dr. Boudewijn Chabot Discuss Assisted Suicide in the Absence of Somatic Illness," *op. cit.*, p. 245.

54 Hendin, "Seduced by Death: Doctors, Patients and the Dutch Cure," *op. cit.*, p. 150.

55 B. Sneiderman and M. Verhoef, "Patient Autonomy and the Defence of Medical Necessity: Five Dutch Euthanasia Cases," *op. cit.*, p. 402.

56 *Ibid.*, p. 403. See also Gene Kaufmann, "State v. Chabot: A Euthanasia Case from the Netherlands," *Ohio Northern University Law Review*, Vol. 20 (1994): 815–820.

punishment or other measure in light of the character of the defendant and the circumstances in which the offence was committed.[57] However, the Amsterdam Medical Disciplinary Court found Dr. Chabot guilty of professional misconduct and imposed the sanction of reprimand. The Disciplinary Court's ruling was based on three accounts: (1) Chabot was faulted for not insisting on therapy as an alternative to assisted suicide. The patient's refusal of treatment should have been a reason for Chabot to refuse the request; (2) Chabot failed to arrange for Mrs. Bosscher to be personally examined by another consultant, a failure which amounted to an ethical breach of duty; and (3) Chabot had not adequately preserved his professional distance, particularly in light of the frequency and length of his sessions with Mrs. Bosscher and the fact that these took place at his home.[58] Dr. Chabot himself regrets his failure to arrange consultation in person with Mrs. Bosscher.[59]

Between 1981 and 1997, there were 20 prosecutions against physicians that ended with a judicial verdict. In nine of these cases, the doctor was found guilty. No punishment was made in three cases and in the other six the doctor was given a conditional sentence without imprisonment. In a few cases, a fine was imposed because the death had been incorrectly reported as natural.[60]

In the United States, Linda Ganzini and colleagues conducted in 1997 a study among board-certified forensic psychiatrists. The study showed that many psychiatrists would support procedural and legal safeguards for patients choosing assisted suicide. For the majority of respondents, a patient requesting assisted suicide would be found competent after an evaluation by two independent examiners, followed by judicial or local administrative review, rendering a determination of competence at a clear and convincing level of proof. The recommended extensive evaluation would assure that only competent patients have access to assisted suicide. The presence of major depression would automatically result in a finding of incompetence.[61]

After laying down the medical and legal foundations for the understanding of the Dutch policy and practice of euthanasia and physician-assisted suicide (PAS), *Part II* elaborates on the main findings of my fieldwork in the Netherlands.

[57] *Nederlandse Jurisprudentie,* 1994, no. 656; John Griffiths, "Assisted Suicide in the Netherlands: The *Chabot* Case," *Modern L. Rev.*, Vol. 58 (March 1995), p. 239.

[58] John Griffiths, "Assisted Suicide in the Netherlands: Postscript to *Chabot,*" *Modern L. Rev.*, Vol. 58 (November 1995), p. 896. See also http://www.bz.minbuza.nl/English/Policy/c_eutheng-A.htm

[59] "Arlene Judith Klotzko and Dr. Boudewijn Chabot Discuss Assisted Suicide in the Absence of Somatic Illness," *op. cit.*, p. 246.

[60] Simon Chesterman, "Last Rights: Euthanasia, the Sanctity of Life, and the Law in the Netherlands and the Northern Territory of Australia," *International and Comparative Law Quarterly,* Vol. 47 (April 1998), pp. 377–378.

[61] Linda Ganzini, Gregory Leong, Darien S. Fenn *et al,* "Evaluation of Competence to Consent to Assisted Suicide: Views of Forensic Psychiatrists," *American J. of Psychiatry,* Vol. 157, No. 4 (April 2000): 595–600, esp. at 599.

PART B

FIELDWORK

CHAPTER 3

THE METHODOLOGY

Before arriving in the Netherlands, I wrote to some distinguished experts in their respective fields: medicine, psychiatry, philosophy, law, social sciences and ethics, asking to meet with them in order to discuss the Dutch policy and practice of euthanasia. Three people advised me that they will be out of the country during the time of my visit. Only one – Dr. Chabot – explicitly declined my request for an interview, saying that he preferred to stay in the background rather than be interviewed.[1] The majority of interviewees were known to me through their writings. The remainder of interviewees was suggested to me by colleagues.

PHASE I – THE INTERVIEWS

The interviews took place during July–August 1999 in the Netherlands. One interview, with Bert Keizer, was conducted in April 2002. Prior to each interview I told the interviewee that the interview is conducted as part of my research on euthanasia in the Netherlands, that I intend to use the material compiled during the interview for my research, and that I will send him/her the content of the interview prior publication. The interviews lasted between one to three hours each, with most taking more than two hours. During the interviews, I asked more or less the same series of questions[2] and took extensive notes that fill some 200 pages. Later the interviews were typed and analyzed.[3]

The interviews were conducted in English, usually in the interviewees' offices. Four interviews were conducted at the interviewees' private homes, and four interviews took place in "neutral" locations, such as coffee shops and restaurants. Two interviews were conducted at the office kindly made available to me at the Department of Medical Ethics, Free University of Amsterdam. To obtain a sampling from different locations, I traveled from Groningen in the north to Maastricht in the south, making extensive use of the efficient train system in the Netherlands.

The interviews were semi-structured. I began with a list of 15 questions, but did not insist on answers to all of them if I saw that the interviewee preferred to speak about subjects that were not included in the original questionnaire. With a few interviewees – most notably the General Practitioners (GPs) performing euthanasia and the de B. family, who openly

1 In his letter dated June 5, 1999, Dr. Chabot wrote: "After four years waiting for the final court judgement (1991–1995) and discussing the case with many people from abroad, I hope you will understand that I prefer to remain in the background now and not to make an appointment with you." He, however, agreed to answer via e-mail some specific questions relating to his conduct that brought about the charges against him.

2 My questionnaire had 15 questions. The Dutch comprehensive study of 1995 consisted of 120 pages and the interviews lasted for an average of 2.5 hours. Cf. Paul J. van der Maas, Gerrit van der Wal, Ilinka Haverkate *et al.*, "Euthanasia, Physician-Assisted Suicide, and Other Medical Practices Involving the End of Life in the Netherlands, 1990–1995," *New Eng. J. of Med.*, Vol. 335, No. 22 (November 28, 1996), at 1700.

3 I am grateful to UCLA School of Law for the generous assistance.

discussed their own personal experience with euthanasia (the head of family underwent euthanasia) – I spoke only about their direct involvement in the practice of euthanasia. As I was interested in the problematic aspects of the euthanasia practice, after some general questions I addressed the troublesome aspects reiterated in the Remmelink report and in the critical discussions mentioned *supra*. This line of questioning disturbed some of the interviewees, who wanted to know my own opinion on the subject matter before continuing to answer my questions. Others seemed eager to bring the interview to a close.

I was struck by the defensiveness expressed by some of the interviewees. Carlos F. Gomez also reported suspicion and guardedness on the part of his interviewees.[4] The attitude of some of my interviewees reminded me of my own initial reaction when I attended debates of post-Zionists outside of Israel during the late 1980s and early 1990s. At that time, I felt that the "dirty laundry" should not be aired in public, that the debate should be restricted to Israelis who are familiar with the intricate aspects of the debate, and that everyone taking part in open forums should show responsibility when addressing the issue before non-Israelis and non-Jews, who might then exploit the information to harm Israel's interests. In the Netherlands, I sensed that the interviewees did not like the idea of a foreigner asking these questions. Their attitude spurred me to entitle one of the first articles I published as a result of this research AN *OUTSIDER*'S VIEW ON THE DUTCH EUTHANASIA POLICY.[5] Although they realized that their euthanasia policy is imperfect, they tried to defend it to the best of their abilities.[6] As a matter of fact, I was somewhat troubled by their lack of criticism and their readiness to accept the euthanasia policy and practice with all of the accompanying flaws.[7] I presume that some of the interviewees identify with their government's decision-making to the extent of defending the system and suspecting foreigners like me who press them with difficult questions. I also suspect that after the publications of Gomez,[8] Keown,[9] and Hendin,[10] they were not enthusiastic about cooperating with me. One interviewee was candid enough to tell me this directly. When I asked why he was willing to sit with me and answer my questions, he replied that he felt obliged as a researcher and scientist to cooperate and wanted his viewpoint to be heard.

4 Carlos F. Gomez, *Regulating Death, op. cit.*, pp. 59–60.

5 *Issues in Law and Medicine*, Vol. 17, No. 1 (Summer 2001): 35–68.

6 In her remarks on the first draft of this study, Heleen Dupuis wrote: "We do not want to defend our views, nor do we want to persuade others to adopt them. We are just very weary when the hundred and umpteenth foreigners come with questions we already have discussed the same number of times. Personally I am very tired by the endless interrogations, whereas I feel that euthanasia is a private matter, such as abortion, and even more so. I also feel that there is a certain exaggeration when it comes to the gravity of the problem." Personal communication on July 25, 2000.

7 In his comments on the first draft of this study, Henk Leenen wrote that he doesn't agree that there is a lack of criticism in the Netherlands: "We have for more than 25 years discussed euthanasia publicly and between all kinds of opinions in a good atmosphere. Nobody was excluded. I personally lectured in meetings of opponents who invited me. I don't know of a country where this is possible." Leenen maintained that gradually a kind of consensus has grown "within a majority" and the problem is that "people like Fenigsen" never took part in this debate and only ventilated their opinions elsewhere. Letter dated July 25, 2000.

8 Carlos F. Gomez, *Regulating Death, op. cit.*

9 John Keown, "The Law and Practice of Euthanasia in the Netherlands," *The Law Quarterly Review*, Vol. 108 (January 1992): 51–78; *idem*, "Euthanasia in the Netherlands: Sliding Down the Slippery Slope?," *Notre Dame J. of Law, Ethics and Public Policy*, Vol. 9 (1995): 407–448.

10 Herbert Hendin, *Seduced by Death* (New York: W.W. Norton, 1997).

Some of the interviewees were nominated by the Dutch government to conduct research on the policy and practice of euthanasia and to submit their recommendations for changes. Science commissioned by the state can be a tricky issue. The researcher might become identified with the project to the extent of becoming "the voice of the state" and forgoing impartiality. It is preferable that research on controversial matters be funded by non-partisan foundations, rather than by an interested government.[11]

Appendix I compiles the list of interviews held during the summer of 1999, including the place and date of each interview.

PHASE II – THE INTERVIEWEES' GENERAL COMMENTS

Prior to each interview I pledged to my interviewees that I would send them the rough draft of the entire manuscript prior to submitting the study for publication. After completing the first draft of writing in July 2000, I sent it to all the interviewees, inviting their comments and criticisms. In my cover letter, I explained that I wished to give each interviewee an opportunity to see that the references to our discussion adequately represented his or her views. I added that the issue at hand was not my analysis and interpretation. Rather, the aim was to ascertain that the interviewee's views were characterized in a fair and honest manner, and that the opinions attributed to him/her were correct.

Yet two of the interviewees strongly objected to my interpretations of the Dutch policy and practice of mercy killing to the extent that they asked to remove any association to them from the book. Fortunately, the majority of interviewees had more respect for free expression and academic freedom and they commented in writing on the first draft. Those commenting on the draft included Arie J.G. van der Arend, Rob Houtepen, Evert van Leeuwen, Henk Jochemsen, H.J.J. Leenen, Heleen Dupuis, Johannes JM van Delden, John Griffiths, A. van Dantzig, James Kennedy, Ron Berghmans, Ruud ter Meulen, Govert den Hartogh, Paul van der Maas and Gerrit Kimsma. Arko Oderwald provided some comments. George Beusmans and Bert Thijs read the draft and had no problems with my accounts of their views. Chris Rutenfrans has asked to comment on the last draft and promptly provided his concise comments. I also sent the parts concerning the *Chabot* case to Boudewijn Chabot, who provided thorough comments and further material about his court case. The interviewees' comments on specific points and their arguments during the interviews are integrated into the discussion. Their general comments are assembled in *Phase II – Interviewees' General Comments.*

PHASE III – UPDATES

As said, the initial fieldwork was conducted during the summer of 1999. On April 10, 2001 the euthanasia law was approved by the Dutch Parliament and I thought it is necessary to

[11] In his comments, van der Maas wrote: "Our studies were funded by the Government. There was an independent committee, both in 1990 and 1995, to ensure the scientific quality of the studies. We have performed these studies under the condition that we would report our scientific studies separately from the Remmelink commission report. These studies are examples of how scientific independence should be guaranteed in government funded research." Personal communication on September 18, 2000. Van der Maas does not provide a compelling explanation why he and his colleagues restrained themselves to collecting and presenting the data and refrained from providing critique and suggestions as to how the situation could be improved.

examine my interviewees' initial reaction to the new law. In June 2001 I returned to the Netherlands for two additional weeks. Prior to my arrival I had notified most of the interviewees of my arrival and welcomed them to express their views on two crucial issues: the legislation process and the role of the regional committees.[12]

Seven of the interviewees answered the two queries via electronic or regular mail. Four other interviewees discussed these questions with me over the phone for 10–20 minutes while in the Netherlands. Six of the other interviewees I met in Amsterdam for discussions that lasted from 40 to 120 minutes each. The meetings took place either in the interviewees' offices or in restaurants. During these discussions some of the interviewees raised further issues of concern, which are reported as well. It is striking that those ten interviewees raised more or less the same issues. *Appendix II* compiles the list of interviews and phone calls held during the summer of 2001, including the place and date of each interview.

On April 1, 2002 the euthanasia law was approved by the Dutch Senate, making the Netherlands the first country in the world to legalize mercy killing. I returned to the Netherlands for a final update and met ten interviewees in Amsterdam, Maastricht and The Hague. Nine interviews took place in offices and lasted from one hour to two hours each. The interview with Bert Keizer in a nursing home was conducted in the presence of an IKON TV crew who asked permission to photograph our meeting and exchange. Some months later parts of our conversation were shown on television. Keizer is the only interviewee who joined the project in its final phase, after the withdrawal of one interviewee, and I decided to present his views in *Phase I.* I did not think that the timing of the interview was significant to the extent of presenting his views separately. *Appendix III* compiles the list of interviews held during April 2002.

A personal note before moving to the main part of this book, the interviews. Prior my visit to the Netherlands I supported euthanasia and physician-assisted suicide. I believed that, in certain cases, we should recognize the necessity for ending life. On such occasions, which should be clearly defined, euthanasia and PAS are morally permissible, and I believed that killing and letting die are morally on a par. I published a few articles articulating these views.[13]

[12] This time I contacted only the ethicists, lawyers and psychiatrists whom I thought could offer interesting views on recent developments.

[13] For instance, R. Cohen-Almagor, "Autonomy, Life as an Intrinsic Value, and Death with Dignity," *Science and Engineering Ethics,* Vol. 1, No. 3 (1995): 261–272; "Reflections on the Intriguing Issue of the Right to Die in Dignity," *Israel Law Review,* Vol. 29, No. 4 (1995): 677–701; "The Patients' Right to Die in Dignity and the Role of Their Beloved People," *Annual Review of Law and Ethics*, Vol. 4 (1996): 213–232.

PHASE I: THE INTERVIEWS

CHAPTER 4

WHY THE NETHERLANDS?

The Netherlands was the first democracy to tolerate euthanasia. The opening question was "Why the Netherlands?" Why does the Netherlands accept euthanasia *de facto* if not *de jure*? What are the significant factors that made the Dutch tradition and culture open to the euthanasia movement?

Some literature addresses this question. Bert Gordijn argues that the Dutch policy is a typical example of a policy of pragmatic tolerance. To better understand this societal phenomenon, one should look at Dutch history. The historical roots can be traced to the Dutch republic of the 17th century, where two dominant and sometimes contradictory societal forces influenced Dutch mentality and policy. The merchants advocated peace and freedom through a policy of tolerance, exemptions and compromises. The Calvinists, on the other hand, strove for normative regulation of all human behavior in accordance with their rigid moral standards. These two forces constituted the Dutch policy of pragmatic tolerance: Certain criminal acts remained unpunished, and certain conditions for immunity from criminal prosecution were formulated in advance and in public by the authorities. Gordijn contends that the practice of tolerance of illegal deeds was normatively regulated and that its influence is significant in the present euthanasia policy.[1]

Egbert Schroten as well as John Griffiths and his colleagues provide another layer of explanation. They emphasize the developments that took place since the 1960s, arguing that the 1960s and 1970s were a crucial watershed for Dutch society. From a conservative, tradition-bound country, the Netherlands was transformed into a society of social and cultural experimentation. The Netherlands took a prominent place in the sexual revolution, the legalization of abortion, the acceptance of soft drugs, the democratization of educational institutions, and the questioning of religious authority. The process of secularization that started in the 1960s gradually undermined the status of traditional institutions. In 1958, 24%

[1] Bert Gordijn, "Euthanasie: strafbar und doch zugestanden? Die niederlandische Duldungspolitik in Sachen Euthanasie," *Ethik Med,* Vol. 10 (1998), p. 12.

of the population had no affiliation, 42% belonged to the Roman Catholic Church, and about 33% were affiliated with the Protestant churches. In the 1990s, the respective figures were 57% (no affiliation), 22% (Catholic) and a bit more than 20% (Protestant).[2]

Societal relationships also changed to the effect that the social distance between ordinary people and people in powerful positions declined. As such, ordinary citizens developed expectations about their roles and influence in society and their ability to affect matters concerning their lives. The political elite, in turn, accommodated themselves to the new reality. In many cases, the elite supported the new ideas, and its members were even spokesmen for them. The political culture of conflict-avoidance, as well as the traditional conviction that it is better to guide social developments than to try to stop them, were instrumental in coping with the waves of change.[3]

In a study conducted by Loes Pijnenborg 34 experts (23 from the Netherlands and the rest from the United States, Australia, Canada, the United Kingdom and Germany) were asked two questions: What is the explanation of the fact that the current discussion on euthanasia in the Netherlands differs from the discussion in other countries? Do you think that the current euthanasia practice in the Netherlands differs from that of other countries? According to the respondents, the most important factors influencing the Dutch policy and practice of euthanasia were the relationships among physician, patient and health care (mentioned by 24 respondents); culture and society (22 answers); legal aspects (15); history (11); church and religion (10); definition of euthanasia (6), and moral principles (5).[4]

These issues were reiterated by my interviewees. Some preferred to answer this question by delineating the historical process that brought about the practice of euthanasia.[5] They explained that up until the late 1960s, the Netherlands was a divided society, with Catholic pillars, Protestant pillars, liberal pillars, and socialist pillars. Each organized its own political party, schools, medical organization, newspaper, media, club, union, etc. These pillars, especially the religious ones, eventually broke down and collapsed. The student revolution of 1969 had a profound effect on the country as the social movement rebelled against the traditional values, autocracy and established hierarchies. After more than two decades of digesting the horrors of WWII, the time was ripe for change. The old morality based on religious perceptions had collapsed. Church values were no longer appealing to large sectors of the population, who were looking for a different set of morals upon which to base their lives.

Consequently, the Netherlands had to find a secular basis for morality. One of the essential moral foundations established was that of individual choice, liberation from collective morality, and the shifting of the focus to autonomy and individuality. The religious view of the sacredness of life was abolished and replaced by secular moral views. The discussion on morality has revolved since then around the rights of the individual, including

2 Proceedings of *Euthanasia and Assisted Suicide in the Netherlands and in Europe*, Maastricht, June 10–11, 1994 (Luxembourg: Office for Official Publications of the European Communities, 1996), p. 101.

3 John Griffiths, Alex Bood and Heleen Weyers, *Euthanasia and Law in the Netherlands* (Amsterdam: Amsterdam University Press, 1998), pp. 12–13.

4 Loes Pijnenborg, "The Dutch Euthanasia Debate in International Perspective," in *End-of-Life Decisions in Dutch Medical Practice* (Department of Public Health, Erasmus University, Rotterdam, 1995, doctorate thesis), pp. 119–132. I am most grateful to Paul van der Maas for sending me this work as well as some other publications.

5 James Kennedy, Govert den Hartogh, Heleen Dupuis, Henk Jochemsen, and A. van Dantzig.

issues like contraception, abortion (the law became liberal in 1969), suffering, and ending of life. The increased democratization of society and the emphasis on individual control thus made the practice of euthanasia more open and available.[6]

Henk Jochemsen, Director of the Professor Lindeboom Institute, who also holds a privately funded chair for medical ethics at the Free University, further explained that Calvinism stresses personal responsibility and that there should be no contradiction between teaching and practice. When physicians realized that there were occasions when euthanasia should be performed, they started pressing for some accommodation to utilize this practice.[7]

In 1969, Jan Hendrik van den Berg, a renowned psychiatrist, published his book *Medical Power and Medical Ethics.*[8] The book discussed the tragic conditions of various patients, who were described as victims of medical power. Had they lived 100 or even 50 years ago, they would have been allowed to die in peace. These days, however, they are being kept alive simply because the capacity to do so is available, regardless of the patients' best interests. Van den Berg suggested granting those victims of medical power a dignified death.[9] James Kennedy, Egbert Schroten and H.J.J. Leenen maintain that this book started the euthanasia discussion in earnest, focusing on life that had lost its purpose. The thesis spoke of recognizing the need to put an end to pointless suffering with the help of a doctor.[10] At that time, the medical establishment was very much against physician-assisted suicide and euthanasia. Some journals, including *Tijdschrift voor Geneeskunde* (*Journal of Medicine*), refused to discuss euthanasia for a long time. Another leading medical journal, *Medisch Contact,* was unique in allowing letters and articles to be published on this issue.

In 1972, the Council of Health commissioned its own study on euthanasia, concluding that euthanasia ought to remain illegal, but that public opinion was changing and that there was a need for reprisal.[11] In 1973, the first euthanasia court case occurred in Leeuwarden: the *Postma* case discussed *supra,* which received a great deal of publicity, prompting certain groups to argue that euthanasia should be allowed. Leenen and van der Maas see this court case as most instrumental in paving the way toward tolerating euthanasia.

6 Evert van Leeuwen, Jaap Visser, Bert Thijs, and Gerrit van der Wal expressed similar views.

7 Ruud ter Meulen made a similar point. For further deliberation on Calvinism, see David E. Holwerda (ed.), *Exploring the Heritage of John Calvin* (Grand Rapids, Michigan: Baker Book House, 1976). See especially an article by Theodore Minnema, "Calvin's Interpretation of Human Suffering," pp. 140–162.

8 The book was translated to English and published by W.W. Norton in 1978.

9 Joop van Holsteyn and Margo Trappenburg, "Citizens' Opinions on New Forms of Euthanasia. A Report from the Netherlands," *Patient Education and Counseling*, Vol. 35 (1998), p. 63.

10 Van den Berg's book responded to widely felt concerns and was reprinted 21 times within seven years and thoroughly discussed in magazines and other media. See John Griffiths, Alex Bood and Heleen Weyers, *Euthanasia and Law in the Netherlands, op. cit.*, p. 48.

11 On changes in public opinion from 1966 to 1991, see Paul J. van der Maas, Loes Pijnenborg and Johannes J.M. van Delden, "Changes in Dutch Opinions on Active Euthanasia, 1966 Through 1991," *JAMA,* Vol. 273, No. 18 (May 10, 1995): 1411–1414; Dick J. Hessing, John R. Blad and Roel Pieterman, "Practical Reasons and Reasonable Practice: The Case of Euthanasia in the Netherlands," *J. of Social Issues,* Vol. 52, No. 2 (1996): 161–166. According to two consecutive polls, 70% of the Dutch people accepted active euthanasia in 1985, and 76% in 1986. Cf. Richard Fenigsen, "A Case Against Dutch Euthanasia," *The Hastings Center Report*, Vol. 19, Special Supp., (January/February 1989): 22–25. A 1998 survey showed that more than 90 percent of the population was in favor of euthanasia. Cf. Suzanne Daley, "The Dutch Seek to Legalize Long-Tolerated Euthanasia," *The New York Times* (June 20, 2000). URL:
http://www10.nytimes.com/library/world/europe/062000holland-mercy.html

In the mid-1970s, the Dutch Voluntary Euthanasia Society was formed and the discussion spread into different circles: media, literature, politics, and academia. The Conservatives, Socialists and Social-Liberals all included the issue of euthanasia in their political platforms. Long deliberations in the Parliament failed to result in concrete steps, as all initiatives were blocked by the ruling party, the Christian-Democrats.

Many physicians felt that the issue needed to be addressed carefully and sincerely. The Dutch Medical Association set up a committee in 1983 to examine the practice of euthanasia. The Society did not adopt a specific standpoint, but nevertheless said that if physicians practiced euthanasia, they needed to follow the jurisprudential Guidelines that Henk Leenen helped to formulate. This was a very important step in the process, and Leenen perceives those years as the formative years in the euthanasia debate. A social movement was created, involving a variety of different interests: physicians, patients' groups, politicians, lawyers, courts, and religious organizations.

In 1987, a state committee, under a Catholic president, recommended legalizing euthanasia, thereby according it a legitimate status for the first time. However, the Christian-Democrats continued to block all legislative attempts. Leenen and Kennedy emphasize that during this period, physicians continued to practice euthanasia and more cases were brought before the courts. Arie van der Arend and Govert den Hartogh noted that patients' groups and voluntary euthanasia advocates gained a stronger influence then.

John Griffiths explains that the Dutch have a history of tolerance and moral ambiguity – they tolerate things that are illegal but not undesirable enough to stop. As Paul van der Maas puts it, the Dutch raise moral questions and try to settle them through pragmatic means. For instance, one of the problems faced by the Netherlands in the last quarter of the twentieth century was that of squatters, people who illegally resided in empty buildings. According to Griffiths, the city government of Groningen understood that the problem might become worse if they evacuated the squatters by force. Hence, it tried to find them housing, and the squatters even formed their own organization to negotiate with the local municipalities.

A. van Dantzig, one of the nation's most well-known psychiatrists, pointed out another phenomenon that illustrates how the Dutch have their own way of dealing with moral questions. Their attitude of "live and let live" is exemplified by the policy on soft drugs. It is forbidden for coffee shops to buy drugs, but they are allowed to sell them, and the authorities don't ask how the shops obtained the drugs. In essence, this is a politics of accommodation and compromise between the interests of each fragment of society as long as they recognize the authority of the State.

Not everyone is happy with the increased sense of secularization and the prevailing liberalism. G.F. Koerselman, another well-known psychiatrist, explains that the Netherlands shifted too rapidly from one extreme to another. Whereas before the 1950s the country was very religious, since then the secularization process has brought about "totalitarian humanism." It is totalitarian because humanism cannot be discussed; it is self-evident, unquestioned, taken for granted.

This view has significant implications for public life. First and foremost, there is absolute respect for individual autonomy. It is almost dogma that people should be, and are, autonomous. Consequently, the prevailing view is that decisions, especially on life and death, should be left in the hands of individuals. Second, argues Koerselman, there is a societal battle against suffering. People should not suffer and, in the name of relieving suffering, it is thought better to seek death. Euthanasia is one form of resolving the issue of suffering. Similarly, Henk Jochemsen says that the social climate in the Netherlands is one in which euthanasia seems a good solution for suffering, and that it is up to the doctor to provide this solution.

Interviewees emphasized "openness" as the trait that made euthanasia possible. Dutch people prefer to put things on the table and to discuss everything openly – from drugs, prostitution and pornography to homosexuality, abortion and euthanasia. Sex education is explicit and universal in Dutch schools. There are open debates on moral issues, a plurality of views, and an overall atmosphere of permissiveness.[12] Recently the Netherlands became the first country to allow gay couples to legally marry and adopt children. Indeed, in a comparative survey of 15 countries, including the United States and Canada, the Dutch had by far the most permissive orientation.[13]

Interviewees noted that the Dutch enjoy having the kind of theological discussion that includes probing norms and values. They further mentioned compromise as a basic mechanism by which resolutions are reached, explaining that the Netherlands has been in trade for centuries and consequently developed the necessary ability to compromise. The ambiguity of having euthanasia Guidelines, yet prohibiting euthanasia under the Penal Code, is the result of compromise between the religious and secular circles of society, between the legalistic and the realistic, in trying to resolve the issue of suffering.[14] The Netherlands has a history of consensus governments based on compromise and mutual tolerance. The Dutch resent authority and have never had an authoritarian regime. As Gerrit van der Wal and Johannes JM van Delden have pointed out, the average Dutch citizen is liberal, pragmatic and tolerant, and appreciates plurality. There is a sense of obligation in Dutch society to try to reach a consensus on controversial matters.

Bert Keizer explained that the Dutch are a relatively stable society; "our commonwealth is not under threat"; people dare to address complicated issues such as euthanasia. The Dutch have the time to focus on such issues. Furthermore, the Netherlands is a country of 16 million people, 16 million theologians who ask themselves ethical questions. Keizer maintained: "You ask: Why the Netherlands? I, on the other hand, ask why other countries are so uptight about euthanasia, not to say hypocritical or cowardly."

Another important consideration is the Dutch attitude in regard to the law. Some interviewees argue that one of the lessons of WWII relates to the limits of authority and the need to dissent when strong moral grounds are provided.[15] Many Dutch physicians during

12 Interviews with John Griffiths, J.K. Gevers, Bert Thijs, Arie van der Arend, Ron Berghmans, and Ruud ter Meulen.

13 Peter Ester, Loek Halman and Ruud de Moor, *The Individualizing Society: Value Change in Europe and North America* (Tilburg: Tilburg University Press, 1994), pp. 56–60.

14 Interviews with Jaap Visser and John Griffiths. This point was reiterated also by Frank Koerselman, Egbert Schroten and Govert den Hartogh.

15 Evert van Leeuwen, John Griffiths and Gerrit van der Wal.

the war felt that the state should not intervene in the doctor-patient relationship and thus refused to participate in the German euthanasia program, openly defying an order to treat only those patients who had a good chance of full recovery.[16] In this context, Chris Rutenfrans spoke of a lack of sufficient safeguards against abuse as a result of historical shortsightedness. Unlike the United States, Germany and other countries where the eugenic movement was active during the 1930s, the Netherlands did not have such a movement. While those countries were alarmed at how the Nazis implemented euthanasia, the Dutch collective memory is unaware of the Nazi experience in this regard. If one has compelling moral reasons, it is considered acceptable to break the law in compliance with one's conscience. The fact that the courts tended to be very lenient towards doctors who performed euthanasia reflects the prevailing positive attitude on this issue.

Many interviewees emphasized the importance of the Dutch healthcare system. As previously mentioned, the Netherlands has inclusive health insurance based on general practitioners who have longstanding relationships with their patients. The result is that most patients trust their doctors.[17] Emphasis is put on the autonomy of the patients, enabling them to have self-determination. At the same time, the strong trust in their GPs may lead patients to yield their autonomy or, ultimately, even their lives. This possibility exists especially as long as doctors suggest euthanasia to their patients.[18]

The literature describes the strongly developed system of primary care in the Netherlands, with 6300 general practitioners and additional nursing care at home. Many patients (40%) die at home, especially patients with cancer (48% of all cancer deaths). Almost all patients (99.4%) have health care insurance, and 100% of the population is insured for the cost of protracted illness.[19]

When faced with the choice of going to the hospital to obtain the necessary health care or to receive the care at home, the Dutch generally prefer the latter. The same is also true for giving birth, a high percentage of which take place at home. Although women are aware of the possible risks involved in giving birth at home, many still prefer home delivery. Home is seen as the place to start life and to finish life. The GP typically visits the patient at home and

16 Leo Alexander, "Medical Science Under Dictatorship," *New Eng. J. of Med.*, Vol. 241 (July 14, 1949), p. 45.

17 Interviews with J.K. Gevers, Jaap Visser, Heleen Dupuis, Gerrit van der Wal, Henk Jochemsen, Govert den Hartogh, Arko Oderwald and Rob Houtepen.

18 This is one of the reasons why doctors should not initiate discussions on euthanasia. I will address this issue later on. Leenen contests this view, arguing that a recent study in which he is involved shows that patients do not yield autonomy due to their trust in their GPs. Letter dated July 25, 2000.

19 Gerrit van der Wal and Robert J.M. Dillmann, "Euthanasia in the Netherlands," *British Medical Journal,* Vol. 308 (1994), p. 1346. For further deliberation on the Dutch health care system, see Jan van der Made and Hans Maarse, "Access to Health Care in the Netherlands," in Jo Lenaghan (ed.), *Hard Choices in Health Care: Rights and Rationing in Europe* (London: BMJ Publishing Group, 1997): 93–111.

establishes a personal relationship that is discrete and private. In this realm of intimacy, outside control often does not exist and is conceived by both parties as interfering and damaging to the personal trust and special bond that have evolved over time. Many incidents of euthanasia happen at home and are not reported because this is considered a private matter, something between the patient, his/her family, and the doctor.

The next chapter reports the general views on the practice of euthanasia in the Netherlands.

CHAPTER 5

VIEWS ON THE PRACTICE OF EUTHANASIA

My next question was formulated in general terms. I wondered how the interviewees felt in regard to the practice of euthanasia and whether they were content with the way in which it is conducted. The majority of interviewees expressed some reservations, but at the same time felt that the system worked relatively well and that euthanasia should be permitted. A small minority felt that the system did not work (i.e., that the Guidelines were quite often violated) and that euthanasia should not be allowed. This small minority complained that the establishment tried to silence them, labelled them as reactionaries and/or religious fundamentalists, and dismissed their point of view.

Sjef Gevers and Bert Keizer have a positive attitude toward the policy. In Gever's opinion, every society will have to address the issue and people will want to have a say about what happens at the end of their lives. The difficult challenge is to develop public policy. Gevers and Keizer see that there are risks involved, but that in the Netherlands, these risks are discussed and generally contained and controlled. They feel that there is no abuse and that, on the whole, the policy works and that the practice is fine.

Heleen Dupuis, an ethicist who has written widely on euthanasia, also expresses satisfaction with the euthanasia policy. She argues that very limited groups (e.g., cancer and AIDS patients) request euthanasia and that their lives are shortened by one or two weeks at the most.[1] One of the principal Dutch authors on research in the fields of euthanasia and physician-assisted suicide (PAS) declared "it is pretty safe to die in Holland." Political scientists estimate that 80% and more of the Dutch population are in favor of euthanasia under certain conditions.[2]

Several authorities[3] said that the Dutch euthanasia policy is basically a good policy and that attempting to forbid euthanasia by law is misguided in light of how often it is done. They are satisfied that the option of euthanasia is available. In Rob Houtepen's view, the major problem is the low level of reporting, most of which can be attributed to laziness and can hopefully be rectified by the regional committees. Van Delden maintained that the rules are flexible, and clinical circumstances do not always conform to solid-rock guidelines.

1 In her comments on the first draft of this essay, Dupuis clarified: "What I wanted to say by that is that clearly euthanasia is not about 'life and death,' as you say somewhere in the book, but about a way of dying of patients who will die anyway in a short period of time. Assisted suicide is a different matter, and morally much more complicated." Personal communication on July 25, 2000.

2 Joop van Holsteyn and Margo Trappenburg, *Het laatste oordeel. Meningen over nieuwe vormen van euthanasie* (Ambo, Baarn 1996). Likewise, Govert den Hartogh and Egbert Schroten are satisfied with the existing arrangement. At the same time, they voice their fear that the new legislation might extend the options for euthanasia too far. For discussion on the law proposal, see H.J.J. Leenen, "Bill on Euthanasia and Assisting Suicide in the Netherlands," *European J. of Health Law*, Vol. 5 (1998): 299–324.

3 A. van Dantzig, Rob Houtepen, Johannes van Delden and Henri Wijsbek.

However, there is danger in interpretation: what does "unbearable suffering" mean? At the same time, these authorities voiced their trust in doctors, claiming that neither life-shortening without the patient's explicit request nor administrating mortal doses of morphine without apparent justification happens in the Netherlands more than in other countries.[4] Given that it is difficult to deal with the euthanasia policy, that there are and will always be some pitfalls, much of the practice is out in the open and the situation is quite satisfactory.

Similarly, H.J.J. Leenen is satisfied with the practice of euthanasia, arguing that no slippery slope exists. He emphasizes that very few doctors have been convicted for euthanasia and that there is positive cooperation on this issue among different circles of society. He notes that 88% of the population is in favor of the practice. In his comments on the first draft of this study, he asked to add that he strongly objects to not reporting euthanasia and that he insists on this issue in his law proposal.[5]

Gerrit Kimsma is "fairly happy with the practice," conceiving it as brave and pragmatic. He does, however, have reservations in the area of clinical care, arguing that inadequate pain relief might lead patients to request euthanasia. He maintains that the Netherlands does well in palliative care, as compared to other countries, but still needs to improve its expertise in pain treatment. Kimsma notes that there are no comparative studies on the level and quality of expertise in pain management. He also believes that the Dutch neglect the area of transference and counter-transference,[6] claiming that sometimes physicians proceed with euthanasia because their patients manipulate them. But, Kimsma immediately adds, this can be said in only a minority of cases. On the whole, he thinks that the euthanasia policy represents the results of mediation between two competing values: protection of life vs. alleviation of suffering. This policy has caused a dramatic change in medicine because what used to be hidden is now public and open.

Some interviewees expressed reservations regarding the lack of control mechanisms. John Griffiths reiterates the problem that many physicians practicing euthanasia do not report their conduct because they do not want to be prosecuted or to provide an opportunity for prosecution. He seemed more troubled than Houtepen on this issue but, like Houtepen, Griffiths argued that the situation in the Netherlands is better than in any other place in the world, and that no slippery slope exists.[7] He assumes that the control system will be improved. Similarly, Evert van Leeuwen and Arie van der Arend think that euthanasia has a place in the medical profession, but at the same time are worried about the lack of reporting. In their view, physicians need to understand that death is not only a medical issue, but also a public issue.

Govert den Hartogh argues that in this area there is no social policy without risk or loss. Given the ideology of autonomy prevalent in the Netherlands the risk of the Dutch policy is

4 However, there are studies that show euthanasia is more prevalent in the Netherlands than in other countries. See, for instance, Martien T. Muller, Gerrit K. Kimsma and Gerrit van der Wal, "Euthanasia and Assisted Suicide: Facts, Figures and Fancies with Special Regard to Old Age," *Drugs & Aging,* Vol. 13, No. 3 (September 1998): 185–191.

5 Letter dated July 25, 2000.

6 For discussion on these concepts, see Jay Katz, *The Silent World of Doctor and Patient* (New York: The Free Press, 1984): 142–150.

7 See also John Griffiths, Alex Bood and Heleen Weyers, *Euthanasia and Law in the Netherlands, op. cit.*, pp. 299–305.

that the life of some patients will be ended on their request, even if they could have had a worthwhile span of life. But it must be recognized on the other hand that there are cases in which life is indeed a hell and in which euthanasia constitutes the best solution. Weighing the risks den Hartogh favors continuation of the existing policy, despite the problems involved in the actual practice.

A few of the interviewees, including Koerselman and Jochemsen as well as Rutenfrans, are well known for their critique of the Dutch policy and practice of euthanasia. G.F. Koerselman was described to me as the most outspoken critic of Dutch euthanasia practice in general and assisted suicide with psychiatric patients in particular. He certainly lived up to his reputation by voicing his grave concern about the developments in the field, warning that the Netherlands is on the wrong road. Koerselman works in two hospitals and says that it is common practice to make euthanasia decisions too quickly and without careful attention. As a psychiatrist, he was invited to investigate and speak with patients requesting to die. Occasionally he was called on Friday, and the medical staff urged him "to finish the process" during the weekend. Koerselman maintains that his colleagues stopped using his services as a consultant once they realized that he would not easily authorize euthanasia.

Henk Jochemsen conceives of euthanasia as unethical insofar as the practice runs contrary to a medical ethic that always seeks to sustain life. He feels that the practice is largely out of control and argues that whatever opinion one may hold about euthanasia, the state must be expected to provide effective control. It is impossible to monitor the work of each and every physician, and having such control would necessitate compromising the values of privacy and confidentiality. In essence, he says, effective control on the one hand, and privacy and confidentiality on the other, are mutually exclusive. Furthermore, he believes that the practice of euthanasia is contrary to a situation in which the state is obligated to protect the life of its citizens.[8]

Chris Rutenfrans, a Roman Catholic who declares that his opposition to euthanasia does not stem from his religion but from practical reasoning, also objects to the policy and practice of euthanasia. He contends that euthanasia might have been necessary during the 1920s, but not today, when pain medication is available. He argues that the Dutch have mistakenly generalized from highly exceptional cases, in which painkillers could not help, to a policy of using euthanasia as an easy and acceptable solution for suffering.[9] It is done in the name of self-determination, even when the patient's self-determination is actually being compromised.

It should be noted that in Oregon, where the policy of physician-assisted suicide is closely examined, eight patients in 2000 (29%) and seven patients in 1999 (26%) expressed concern about inadequate control of pain as their illness progressed, as compared with two

8 For further deliberation, see Henk Jochemsen, "The Netherlands Experiment," in John F. Kilner, Arlene B. Miller and Edmund D. Pellegrino (eds.), *Dignity and Dying* (Grand Rapids, MI.: William B. Eerdmans Publishing Co., 1996): 165–179.

9 Van der Maas contests this view. In his comments, he wrote that not only his studies but also studies in other countries have established time and again that pain is rarely the reason for a euthanasia request. Personal communication on September 18, 2000. However, van der Maas's own study from 1990 shows that in 46% of cases, pain was the reason given by the patient for requesting euthanasia or PAS. P.J. van der Maas, J.J.M. van Delden, and L. Pijnenborg, *Euthanasia and other Medical Decisions Concerning the End of Life*, *op. cit.*, pp. 43–44.

patients (12%) in 1998.[10] These findings may reflect advances in palliative care in Oregon, which ranks among the top five states in per capita use of morphine for medical purposes.[11] Other studies have shown that pain is not prominent in oncology patients' attitudes toward PAS. Ezekiel J. Emanuel and colleagues found that patients actually experiencing pain were more likely to find euthanasia or physician-assisted suicide unacceptable.[12]

On the other hand, the Oregon findings that only two patients in 1998 and a more significant number of patients in 1999 and 2000 expressed concern about inadequate pain control may indicate only physicians' opinions that they are capable of managing pain. It might also be the result of poor communication between cancer patients (the majority of patients who asked for lethal drugs) and physicians. This hypothesis is strengthened when one looks at interviews with family members, conducted in 1999. The most frequently cited reasons by family members for the patient's decision to request assistance with death were concern about loss of control of bodily functions (68 percent), loss of autonomy (63 percent), and physical suffering (53 percent).[13] Linda Ganzini and colleagues report that pain was an important consideration for 43 percent of patients who requested prescription for a lethal medication.[14]

Moreover, studies showed that pain control for cancer patients is often inadequate and that physicians typically underestimate pain. Patients with significant pain caused by cancer visit their physicians and frequently leave with as much pain as they came with because their pain was never discussed or treated. Impediments to adequate pain treatment include health care providers' fear of inducing physical or psychological addiction, misconceptions about pain tolerance, and assessment biases.[15] Furthermore, communication about pain often depends on the patients' complaining of it. Patients, however, are often reluctant to report pain for a variety of reasons, including wanting to be a "good" (noncomplaining) patient, concern about having to take strong painkillers, or worries that talking about pain might take too much time and distract the physician from dealing with the disease itself.[16] Moreover, many patients seek pain relief from complementary therapies. Often they feel that these

[10] Amy D. Sullivan, Katrina Hedberg, and David W. Fleming, "Legalized Physician-Assisted Suicide in Oregon – The Second Year," *New Eng. J. of Medicine,* Vol. 342, No. 8 (February 24, 2000), at 600; Amy D. Sullivan *et al.*, "Legalized Physician-Assisted Suicide in Oregon, 1998–2000," *New Eng. J. of Med.*, Vol. 344 (February 22, 2001); Editorial, "The concerns stated by those who received suicide assistance in Oregon call into question why such a practice is needed," *American Medical News* (March 19, 2001) .See also Erin Hoover Barnett, "Oregon's Death With Dignity Act influences end-of-life care across the State as doctors wrestle with prescribing pain treatment," *The Oregonian* (Portland) (February 18, 2001), and http://www.ohd.hr.state.or.us/chs

[11] Arthur E. Chin *et al.*, "Legalized Physician-Assisted Suicide in Oregon: The First Year's Experience," *New Eng. J. of Med.*, Vol. 340, No. 7 (February 18, 1999), p. 582.

[12] Ezekiel J. Emanuel *et al.*, "Euthanasia and Physician-assisted Suicide: Attitudes and Experiences of Oncology Patients, Oncologists, and the Public," *Lancet*, Vol. 347 (June 29, 1996), p. 1809.

[13] Amy D. Sullivan *et al.*, "Legalized Physician-Assisted Suicide in Oregon—The Second Year," *New Eng. J. of Med.*, Vol. 342, No. 8 (February 24, 2000), p. 601.

[14] Linda Ganzini *et al.*, "Physicians' Experiences with the Oregon Death with Dignity Act," *New Eng. J. of Med.*, Vol. 342, No.8 (February 24, 2000), p. 559.

[15] Kathleen Foley, "Dismantling the Barriers: Providing Palliative and Pain Care," *Medical Student Journal of the American Medical Association*, Vol. 283 (January 5, 2000), p. 115.

[16] S. Ward *et al.*, "Patient-Related Barriers to Management of Cancer Pain," *Pain*, Vol. 53 (1993): 319–324; Charles S. Cleeland *et al.*, "Pain and Its Treatment in Outpatients with Metastatic Cancer," *New Eng. J. of Med.*, Vol. 330 (1994): 592–596.

methods offer a holistic approach that is lacking in the traditional allopathic model.[17] It is advisable that doctors examine whether the prescribed pain control is adequate. Palliative care is able to prevent or at least to ease most manifestations of physical pain.[18]

PROS AND CONS OF THE DUTCH EUTHANASIA POLICY AND PRACTICE

This question supplements the former question and is more specific. In response to the former question, the interviewees could have raised any issue that came to mind. This question provided respondents with the opportunity to elaborate on the positive aspects of the Dutch euthanasia policy and practice and, at the same time, pressed those who did not mention the negative aspects to address the problems involved. Many interviewees were eager to speak about the pros and far more reluctant to address the cons, to the extent that I had to press them to answer also the second part of the question. Some were ambivalent about the euthanasia practice. Koerselman, Jochemsen and Rutenfrans who emerged as the critical voices in my study continued to substantiate their grim assessments about the practice.

Heleen Dupuis and A. van Dantzig voiced the most supportive views on the policy and practice of euthanasia.[19] Dupuis, who served as president of the Dutch Voluntary Euthanasia Society, sees only positive aspects in the policy and practice. Dying is a private matter and should be a private choice, with society providing safeguard mechanisms and allowing euthanasia as an option. When pressed about the cons, she responded by saying: "Theoretically there are cons, but I never heard about them."

Van Dantzig emphasizes the openness of the practice, maintaining that it is rational for one to say that upon reaching a certain level of dementia life should be ended. The role of the doctor is to verify that the person has indeed reached that stage. He argues that when a doctor ends a person's life, it is not a crime provided that the act was done on medical grounds. Van Dantzig thinks that it is wrong to criminalize euthanasia and believes that a doctor does not commit a crime when he provides this service. He recommends that physicians interview every person who requests euthanasia. Many times it is possible to find another solution. In any event, van Dantzig claims, someone who wants to carry out euthanasia will do so regardless of law and policy. He says that after accepting the rationale for euthanasia, all that

17 Catherine S. Magid, "Pain, Suffering, and Meaning," *Medical Student Journal of the American Medical Association*, Vol. 283 (January 5, 2000), p. 114.

18 To institute effective pain control, new programs for the training and certification of palliative care consultants need to be developed and implemented. See Franklin G. Miller *et al.*, "Regulating Physician-Assisted Death," *New Eng. J. of Med.*, Vol. 331, No. 2 (July 14, 1994): 119–123; Timothy E. Quill *et al.*, "Palliative Options of Last Resort," *Journal of the American Medical Association*, Vol. 278, No. 23 (December 17, 1997): 2099–2104; P. D. Doyle *et al.* (eds.), *Textbook of Palliative Medicine* (New York: Oxford University Press, 1998); Anne Scott, "Autonomy, Power, and Control in Palliative Care," *Cambridge Quarterly of Healthcare Ethics*, Vol. 8, No. 2 (1999): 139–147; Janet L. Abrahm, "The Role of the Clinician in Palliative Medicine," *Medical Student Journal of the American Medical Association*, Vol. 283 (January 5, 2000), p. 116. For further deliberation, see several articles published in *JAMA*, Vol. 290, No. 18 (November 12, 2003), and http://www.painandthelaw.org/

19 In his comments, van Dantzig wrote: "I have different opinions from Prof. Dupuis. I do not believe in the autonomous decision of the patient, as she does." He noted that his views "are not as radical as you think." Personal communication on July 6, 2000.

is left is to solve the practical problems, and these can be sorted out. Van Dantzig trusts doctors to carry out their professional responsibilities.[20]

Henri Wijsbek was happy to speak about the pros and far less eager to discuss the cons. Wijsbek praised the Guidelines because they insist on a voluntary request by the patient, they attempt to find an answer to the suffering, and they also require consultation. He believes that euthanasia is justified when good reasons are provided. No cons were mentioned.

Paul van der Maas points to a number of relevant issues: the Dutch universal health care coverage that removes the economic pressure of forcing families to spend their resources on their loved ones; the fact that GPs know their patients for many years; and the legal aspect that discourages prosecution if there is ample justification. According to van der Maas, most people feel that there are sufficient safeguards and that the prescribed limits are acceptable and safe, in line with what people want. He also notes that euthanasia is rarely performed – once every 50 deaths – and that it is performed only when people are really suffering.[21] Finally, van der Maas stresses the self-limiting aspects of euthanasia: Most doctors want to avoid euthanasia because they find it too emotional.

Regarding the cons involved in the practice, van der Maas said that the number of requests for euthanasia rose substantially from 1990 to 1995,[22] and one might think that some doctors may be pressured by their patients' requests for euthanasia. Van der Maas also regards it as important that doctors spend sufficient time discussing euthanasia with patients, meaning more than one session. More research should be conducted on this issue.

Bert Keizer who thinks the Dutch practice of euthaansia is fine argues that the pros are that people in unbearable situations can ask for death. Keizer testifies that he supports and performs euthanasia because he hopes that he would receive a similar treatment if required and he would wish it. Keizer finds the term "unbearable suffering" difficult to explain. He says that suffering almost never has to do with pain alone. For instance, when people are old, recognizing that they are unable to escape a certain situation, then they might lose hope, feeling trapped in a situation, and then they might opt for euthanasia as an escape.

The cons are, says Keizer, that once euthanasia is allowed in certain situations, the policy might be stretched too much. The slippery slope is an actual concern. Euthanasia is an ethical issue, and it is the characteristic of an ethical issue that one doesn't know one's way about. Doubts are creeping in, and people may see euthanasia as a solution in less clear cases. Furthermore, the Netherlands is in trouble internationally because of the euthanasia policy. It is criticized by many countries. Some of the criticisms contain blatant lies. Baseless accusations are heard. For instance, argues Keizer, the film *Death on Request* caused the Netherlands a lot of harm.[23] People outside the country criticized the physician's conduct,

[20] In his comments on the first draft of this study, van Dantzig added a general remark: "euthanasia is not killing, it is ending a life. Killing has criminal associations, and should not be used for the professional actions of a doctor." Personal communication on July 6, 2000.

[21] Cf. Gerrit van der Wal and P.J. van der Maas, "Empirical Research on Euthanasia and Other Medical End-of-Life Decisions and the Euthanasia Notification Procedure," in David C. Thomasma *et al.* (eds.), *Asking to Die* (Dordrecht: Kluwer Academic Publishers, 1998), p. 155.

[22] In 1995, the annual number of requests made to doctors for euthanasia or PAS "at a later time" was estimated at 34,500, an increase of 37% as compared with 1990. In total, 9,700 explicit requests were made for euthanasia or PAS, an increase of 9%. Gerrit van der Wal and P.J. van der Maas, "Empirical Research on Euthanasia and Other Medical End-of-Life Decisions and the Euthanasia Notification Procedure," *Ibid.*, p. 155.

[23] Some problematic aspects of this film are discussed in *Chapter 9*.

saying that he lacked the needed expertise to take care of the patient. Keizer does not agree with this criticism.

James Kennedy admits that he is ambivalent about the practice. The positive thing about it is that large segments of society participate in the open and long deliberations. Furthermore, there is sensitivity to suffering and respect for autonomy, as well as a genuine attempt to reach a consensus. Having said that, Kennedy feels that despite its openness, the Dutch public really does not know a lot about the practice of euthanasia. Unlike the specialists, the public is not aware of all the problematic aspects revealed in the 1990 Remmelink report and in the 1995 report, the research on withdrawal and withholding of treatment, or the employment of the double effect doctrine. These issues should be further discussed and researched.

In regard to critics of euthanasia, Kennedy contends that they are not taken seriously enough in the Netherlands. The Dutch are sensitive to criticism from abroad and are defensive about their policies. As Koerselman, Rutenfrans and Jochemsen have testified, Dutch critics are also resented. The common argument against foreign critics is that they do not understand the Dutch policy, whereas the common argument against Dutch critics is that they belong to religious sects.

The Dutch may be overconfident about the need for euthanasia and the justification of its practice. One may assume that if the given Guidelines are not complied with, then people will see that there is a serious problem. Yet, the Dutch people do not seem to feel this way.[24] Indeed, the moral superiority of Dutch euthanasia advocates is evident in Herbert Cohen's claim that they see foreign criticism as evidence that the United States and the rest of the world are not as enlightened as the Netherlands.[25]

Egbert Schroten answers the question in general terms, holding that there is no 100% black or white picture in ethics. There are both pros and cons in the euthanasia policy and practice but, generally speaking, the pros outweigh the cons. Whereas Schroten considers the Guidelines strict and sufficient, van Delden feels that the Guidelines could be more specific and should clarify such vague terms as "no prospect of improvement" and "unbearable suffering." Van Delden adds that an unfortunate result of such ambiguity might be that some patients would be rejected due to their failure to meet the criteria for receiving euthanasia.

Evert van Leeuwen and John Griffiths reiterate the problem of lack of reporting. On the positive side, van Leeuwen speaks of allowing people to die when they are conscious of their surroundings, when they are still able to appreciate their lives, and when they can choose the time of their death. He emphasizes that 90% of the reported cases are terminal cancer patients and 10% are ALS patients, who suffer degenerative diseases – all of them patients who are suffering without hope of improvement or cure.

Ruud ter Meulen and Ron Berghmans stress that the positive aspects of euthanasia practice have to do with the ability to conduct an open discussion and that patients are allowed to play an important role in deciding their fate. Berghmans is in favor of providing assistance in dying and considers it humane to have the option. He maintains that in contrast to Germany, where the climate is not conducive to having a rational and productive debate on euthanasia, in the Netherlands it is possible to discuss everything.

24 In his comments on the first draft of this study, Leenen wrote that the Guidelines are working. Only the procedural requirement of reporting is still defective. Letter dated July 25, 2000.

25 Herbert Hendin, *Seduced by Death* (New York: W.W. Norton, 1997), p. 148.

As for the negative aspects, ter Meulen sees a dangerous shift from allowing euthanasia for cancer patients to allowing it for psychiatric patients. The criteria are in flux, shifting too broadly towards what some might call a slippery slope. The Netherlands started the process with cancer patients and then expanded euthanasia practice to other groups of patients, including psychiatric patients, sick children, and cases of dementia. First, there was insistence on respecting the autonomy of the patient and honoring his/her desire to determine the moment of death. Now the discussion is shifting to mercy killings without the explicit request of the patient. Koerselman, Rutenfrans and Jochemsen have voiced similar concerns.

Ron Berghmans points to the significant number of cases of non-voluntary euthanasia, emphasizing the need to seek more information about these cases. Moreover, like Frank Koerselman, he feels that there is a need to further discuss the issue of advance directives,[26] particularly with regard to dementia patients. The main problem in cases of dementia is in determining how to evaluate suffering, because the regulations require meeting the condition of unbearable suffering. In addition, there is the issue of control. Berghmans fears that this will always remain a problem in regulating euthanasia. He asserts that the Netherlands should strive for 100% transparency, but he is pessimistic about the probability that this will happen. As long as euthanasia remains criminal, Berghmans does not think that the level of doctors' reporting will increase. At the same time, he does not have a clear position on whether euthanasia should be decriminalized.[27]

Chris Rutenfrans focuses only on the cons, arguing that there are many cases of people who are killed prematurely. He claims that euthanasia policy has shifted from voluntary to involuntary, from terminal illness to earlier stages of the disease. Patients are killed when the prospects for health improvement are slim.[28]

26 An advance directive (AD) is a document that allows patients to express what life-sustaining treatments they want and for whom they want to make these decisions for them. In the United States, more than forty states have enacted legislation supporting the use of ADs. For further discussion, see Joseph J. Fins, "The Patient Self-determination Act and Patient-Physician Collaboration in New York State," *N.Y. State J. of Medicine*, Vol. 92 (November 1992): 489–493; Nitsa Kohut and Peter A. Singer, "Advance Directives in Family Practice," *Canadian Family Physician*, Vol. 39 (May 1993): 1087–1093; Maarthen Reinders and Peter A. Singer, "Which Advance Directive Do Patients Prefer?," *Journal of General Internal Medicine*, Vol. 9 (January 1994): 49–51; Dallas M. High, "Families' Roles in Advance Directives," *Hastings Center Report,* Special Supplement (November–December 1994): S16–S18; Stuart Hornett, "Advance Directives: A Legal and Ethical Analysis," in John Keown (ed.), *Euthanasia Examined* (N.Y.: Cambridge University Press, 1995), pp. 297–314; Hans-Martin Sass, Robert M. Veatch and Rihito Kimura (eds.), *Advance Directives and Surrogate Decision Making in Health Care* (Baltimore: Johns Hopkins University Press, 1998); Lawrence P. Ulrich, *The Patient Self-Determination Act* (Washington, D.C.: Georgetown University Press, 1999): 219–251; David Degarzia, "Advance Directives, Dementia, and 'The Someone Else Problem'," *Bioethics,* Vol. 13, No. 5 (1999): 373–391; D. William Molloy, Gordon H. Guyatt, Rosalie Russo *et al,* "Systematic Implementation of an Advance Directive Program in Nursing Homes," *JAMA,* Vol. 283, No. 11 (March 15, 2000); Joan M. Teno, "Advanced Directives for Nursing Home Residents," *JAMA,* Vol. 283, No. 11 (March 15, 2000); Paul Biegler, Cameron Stewart, Julian Savulescu and Loane Skene, "Determining the Validity of Advance Directives," *Medical J. of Australia*, Vol. 172 (2000): 545–548. URL: http://www.mja.com.au/public/issues/172_11_050600/biegler/biegler.html

27 Henk Leenen is similarly happy that Dutch society discusses controversial matters, like euthanasia, in the open. However, he reiterates the need for better control, stressing that euthanasia is not a matter for doctors and patients alone. The entire society needs to get involved in the process, and then the level of reporting should improve.

28 In his comments, van der Maas asks: "Is there any serious research on which Mr. Rutenfrans can establish these claims? I have never seen any." He maintains that he fails to understand how I am using the results of his extensive research, claiming that they may not be impartial because the government funded those studies, and at

Van der Arend, a nurse, is not satisfied with the role of nurses in the euthanasia procedure. Nurses are not systematically involved in the decision about whether to honor euthanasia requests. Furthermore, nurses perform euthanasia in 20% of the cases in hospitals, in clear violation of the Guidelines, which require a doctor to perform euthanasia. According to van der Arend, nurses are often unaware that they are administering euthanasia in violation of the Guidelines. When euthanasia is conducted at patients' homes, nurses are often not consulted about the decision and sometimes discover it only after the death.[29] It should be added that research shows that approximately half of the GPs did not consult with nurses about a patient's request for either euthanasia or assisted suicide, the intention to administer them, and the actual administration.[30]

the same time use "the kind of undocumented insinuations by Mr. Rutenfrans." Personal communication on September 18, 2000. I believe that researchers should better review all possible sources of information.

29 Van der Maas and colleagues report that a nurse or someone else sometimes performs euthanasia with a drug that was prescribed for this purpose by a physician. P.J. van der Maas, J.J.M. van Delden, and L. Pijnenborg, *Euthanasia and other Medical Decisions Concerning the End of Life*, *op. cit.*, p. 193.

30 Martien Tom Muller, *Death on Request* (Amsterdam, Vrije Universiteit Thesis, 1996), p. 80.

CHAPTER 6

WORRISOME DATA

HASTENING DEATH WITHOUT THE EXPLICIT REQUEST OF PATIENTS

The following question opened the critical line of the interviews: “Some of the most worrisome data in the two Dutch studies are concerned with the hastening of death without the explicit request of patients. There were 1000 cases (0.8%) without explicit and persistent request in 1990, and 900 cases (0.7%) in 1995. What is your opinion?”

Most of the interviewees offered similar interpretations of this finding. They said that this group includes cancer patients, Persistent Vegetative State patients,[1] newborns with severe health problems, and patients who are suffering that would die within a matter of days or even hours. By so doing, physicians strive to alleviate the pain of very sick patients at the end of their lives. In essence, what they are saying is that it is right to replace autonomy with beneficence in such severe circumstances. Some of the interviewees saw no problem in this, arguing that the balance favors termination of life in such instances. Many interviewees also emphasized that this practice was common long before euthanasia became available in the Netherlands and that it is not a specifically Dutch problem; there is a similar policy in hospitals all over the world.[2]

Consider the view of one of the foremost euthanasia activists, H.J.J. Leenen, who argues that these are not cases of euthanasia. Indeed, strictly speaking, these are not euthanasia cases because the Dutch definition speaks of patients who made a persistent voluntary request for the practice. Leenen argues that nearly all of the cases involved cancer patients in the last phase of their lives, who were suffering greatly and who had had so much medication that they were no longer competent. Their physicians sometimes help these suffering, dying patients with the last push. Such cases do not indicate a slippery slope, and they occur in every country. Nevertheless, Leenen is worried that the physicians did not discuss the option of termination of treatment early

1 The term PVS is used in reference to patients who are in a twilight zone between life and death. Those patients have suffered brain injuries from different causes: trauma, cerebral anoxia from hypotension or cardiac arrest, cerebrovascular accidents, or dementia. They have periods of wakefulness and physiological sleep/wake cycles, but at no time is the patient aware of himself or herself or the environment. Neurologically, being awake but unaware is the result of a functioning brain stem and the total loss of cerebral cortical functioning. No voluntary action or behavior of any kind is present. On this group of patients, and my ethical objections to the term PVS, see R. Cohen-Almagor, “Some Observations on Post-Coma Unawareness Patients and on Other Forms of Unconscious Patients: Policy Proposals,” *Medicine and Law*, Vol. 16, No. 3 (1997): 451–471; “Language and Reality at the End of Life,” *Journal of Law, Medicine and Ethics*, Vol. 28, No. 3 (Fall 2000): 267–278, and *The Right to Die with Dignity: An Argument in Ethics, Medicine, and Law* (Piscataway, NJ.: Rutgers University Press, 2001). For further deliberation, see Carl Zimmer, "What if there is something going on in there?," *New York Times* (September 28, 2003).

2 Interviews with John Griffiths, Evert van Leeuwen, Govert den Hartogh, Bert Thijs, A. van Dantzig, Heleen Dupuis, Henri Wijsbek, and Arie van der Arend.

on with these patients. Physicians know the consequences of increased doses of medication and should discuss the options at the beginning of the dying process. It is a physician's responsibility to verify what patients want.[3]

Bert Thijs and Evert van Leeuwen recommend ethical and practical training as well as open discussions to overcome this problem. Physicians should discuss and debate this issue by holding regular consultation sessions. These discussions would make it possible for physicians to reflect on their decision-making process among themselves and with paramedics in their teams.

Two of the most productive researchers in the field of euthanasia expressed remarkably similar opinions. They said that they are not worried about the data, explaining that this category of patients, who did not make an explicit request for euthanasia, includes comatose patients whose families asked whether it was necessary to keep the patients alive and who wanted to stop their suffering. Two-thirds were cancer patients in the advanced stage of the disease. They were suffering extremely, vomiting, unable to express themselves, and totally immersed in their agonizing condition. In other countries, Gerrit van der Wal maintains, ending a patient's life in such cases would be called double effect. In the Netherlands, "we are more explicit, more Calvinistic" (I asked what this meant, and he answered "more open, rigid, honest, straightforward"). He further argued that in most of these cases, the issue of euthanasia was not discussed sufficiently beforehand. Therefore, it is important to make the climate around euthanasia more open and to exchange views between physicians and patients. Very few of these patients had living wills, but about half had indicated in the past that they would prefer to hasten death under such terrible conditions; the other half had given no indication of what they would have wanted.[4]

Van der Wal also notes that the doctors' intentions regarding these patients were unclear. He, Heleen Dupuis and another senior researcher argue that most of the patients died from large doses of painkilling morphine ("double effect"), rather than the euthanasia drugs that paralyze the muscles. The doctors wanted to alleviate the pain and suffering of their patients, most of whom were in the advanced stages of cancer. They had only days or hours left to live and had been rendered incompetent from painkilling drugs. The doctors were then forced to

[3] Cf. Jan H. Veldink, John H.J. Wokke, Gerrit van der Wal, J.M.B. Vianney de Jong and Leonard H. van den Berg, "Euthanasia and Physician-assisted Suicide among Patients with Amyotrophic Lateral Sclerosis in the Netherlands," *New Engl. J. Med.*, Vol. 346, No. 21 (May 23, 2002): 1638–1644.

[4] See Herbert Hendin, *Seduced by Death* (New York: W.W. Norton, 1997), pp. 88–89. Jaap Visser argues in answering this question that in many of these cases, the patients had living wills. In contrast, van Delden claimed that a very small number of these patients had a living will. The 1990 study reports that about one-quarter of the patients had previously "indicated something" regarding life termination. Cf. P.J. van der Maas, J.J.M. van Delden, and L. Pijnenborg, *Euthanasia and other Medical Decisions Concerning the End of Life*, Health Policy Monographs (Amsterdam: Elsevier, 1992), p. 69. Van der Wal, van der Maas *et al.* report that in about half of these cases, either the decision was discussed with the patient earlier in the illness or the patient had expressed a wish for euthanasia if suffering became unbearable. In the other cases, the patient was deemed incompetent. Cf. Paul J. van der Maas, Gerrit van der Wal, Ilinka Haverkate *et al.*, "Euthanasia, Physician-Assisted Suicide, and Other Medical Practices Involving the End of Life in the Netherlands, 1990–1995," *New Eng. J. of Med.*, Vol. 335, No. 22 (November 28, 1996), p. 1701. John Griffiths says that this category of patients is highly heterogeneous, including severely defective newborn babies, long-term coma patients, and persons who at some earlier time have expressed a general wish for euthanasia but who in the final stages of the process of dying are no longer capable of expressing their will. John Griffiths, "The Slippery Slope: Are the Dutch Sliding Down or Are They Clambering Up?," in David C. Thomasma *et al.* (eds.), *Asking to Die, op. cit.*, p. 98.

act based on their own initiative without having an opportunity to consult the patients regarding euthanasia.[5]

Bert Keizer asserts that from the British hospice world we hear that patients in the Netherlands are being put to death all the time, "while they in fact treat their patients the way we treat ours by administering liberal dosages of morphine to terminal patients." Doctors resort to morphine to ease suffering. Keizer said that morphine is not administered without asking the consent of patients, or without telling the patients that morphine might shorten life. The truth is you cannot always be sure whether the administered morphine only alleviates pain, while not shortening the patient's life. What happens in the Netherlands happens also in other parts of the world. The Dutch are simply more open and candid about the practice. So the worrisome data do not worry Keizer "at all".

Johannes van Delden provides the most interesting answer to the question regarding these worrisome data. He said that these findings were in part artifact, the creation of the researchers in the construction of the categories, which included clear-cut expressed wishes, cases of no explicit request, and cases with some remarks. Of the 1000, 56% expressed some view. However, the authors of the research "decided to have a clear-cut euthanasia group, and this group emerged from the way we chose to analyze." Van Delden explained, "You need autonomy and beneficence to perform euthanasia." There might be *very* extreme cases in which doctors decide to conduct euthanasia without the patient's explicit request, but these cases are few in number. It is hardly conceivable to act in this way, "maybe in pediatrics." Van Delden did recognize that most of these cases were insufficiently justified and insisted that the autonomy requirement be maintained.[6]

Henri Wijsbek, Arie van der Arend and Rob Houtepen see no reason for alarm in regard to this data. Most of these cases involve situations in which patients are unconscious, suffering greatly, in the last stage of disease, and physicians then do not have any other choice but to end the patient's life. Wijsbek says that those patients would have died within a short time.[7] This practice occurs in many countries, and Wijsbek sees no problem with it. He adds his hope that under such circumstances, his doctors would do the same for him. When there is hopeless suffering, and doctors cannot alleviate the pain, the option of euthanasia should be available. Van der Arend is not worried about the numbers, saying that he could

5 For further deliberation, see Gerrit van der Wal, "Unrequested Termination of Life: Is It Permissible?," *Bioethics*, Vol. 7, No. 4 (1993): 330–339.

6 In another forum, van Delden explained that the type of patients involved were cancer patients with brain metastases and, consequently, were rendered incompetent. It would give "a false picture of Holland if one thought that we put away demented people or the weak ones in society. It really is the patient who is dying and already fading away." Proceedings of *Euthanasia and Assisted Suicide in the Netherlands and in Europe*, Maastricht, June 10–11, 1994 (Luxembourg: Office for Official Publications of the European Communities, 1996), p. 47.

7 According to van der Wal, van der Maas *et al.*, life was shortened by 24 hours at most in 33% of these cases, and in a further 58% it was shortened by one week at most. Cf. Paul J. van der Maas, Gerrit van der Wal, Ilinka Haverkate *et al.*, "Euthanasia, Physician-Assisted Suicide, and Other Medical Practices Involving the End of Life in the Netherlands, 1990–1995," *op. cit.*, p. 1702. The 1990 study reports that in 21% of the cases, life was shortened by one to four weeks; in 7% of the cases, life was shortened by one to six months; and in a small number of cases, life was shortened by more than half a year. These patients were not in the terminal stage of their illness. Cf. P.J. van der Maas, J.J.M. van Delden, and L. Pijnenborg, *Euthanasia and other Medical Decisions Concerning the End of Life*, *op. cit.*, pp. 66, 183. See also Loes Pijnenborg, Paul J. van der Maas, J.J.M. van Delden and Caspar W.N. Looman, "Life-terminating Acts without Explicit Request of Patient," *Lancet*, Vol. 341 (May 8, 1993): 1196–1199.

not imagine that termination of life was the result of careless medical practice. He added: "There will always be cases like these."

In turn, Rob Houtepen testifies that he is quite liberal about termination of life when people are suffering, even if they are incompetent. He believes that compassion is the primary consideration for euthanasia. Autonomy is a secondary consideration. It is unjust that people be denied the option to end their suffering, an option that is available to competent patients. In his mind, we should not make strict distinctions between competent and incompetent patients. Hence, Houtepen is "not shocked" by the figures, though he does recognize the need for stricter notification procedures so as to make more data available about the circumstances of each and every patient who did not make an explicit request.

Ron Berghmans supports the need for more information about these cases and the circumstances involved in each case, particularly when the patient could have expressed an opinion about the issue. In addition, the issue of advance directives (ADs) needs to be discussed further. The new law stipulates that ADs have the legal force of the patient's voluntary explicit request. Berghmans reiterates the problem of how to evaluate suffering in cases of dementia and expresses doubt about whether we should honor the ADs of dementia patients.[8]

In his comments on the first draft of this study, Govert den Hartogh wrote that my implicit assumption is that the worrisome data are the result of the lax Dutch rules. He thinks that probably the opposite is true: In other countries, this figure would probably be much higher, precisely because euthanasia is forbidden and hence cannot safely be discussed with patients. Accordingly, the result of the Dutch rules, on this account, is that the figure is unusually low. This, however, does not mean that the data are not worrisome.

To substantiate his point, den Hartogh referred to the recent research project in Flandres, which is a replica of the van der Maas/van der Wal studies. Belgian law at that time forbade euthanasia and assisted suicide, and there is no equivalent there to the Dutch judge-made law. According to this research, the overall percentage of deaths as a result of euthanasia or assisted suicide in Belgium is somewhat lower, but not very much lower, than it is in the Netherlands. However, the percentage of deaths from "euthanasia" without explicit request is five times as high (3.5%).[9] Den Hartogh's interpretation of these data is not that Belgian doctors, any more than Dutch doctors, "get rid of unwanted patients." Although they believe themselves to be acting in the best interests of the patient and fulfilling the wishes of the patient, they are simply reluctant to discuss the matter openly.[10] It is outside the scope of this discussion to delve into an extensive review of the Belgian situation. I should say, however, that my independent research in Belgium which included survey of literature and interviews with some leading experts reveals that the policy and practice of euthanasia in Belgium is worrisome, and that further investigations are warranted to shed light on the scope of abuse.

A minority of interviewees did express concern in regard to these data. When patients are suffering and on the verge of death, they do not view it as problematic if the doctor provides

[8] For further deliberation, see R. Berghmans, "Advance Directives and Dementia," in R. Cohen-Almagor (ed.), *Medical Ethics at the Dawn of the 21st Century* (New York: New York Academy of Sciences, 2000), Vol. 913 of the *Annals*: 105–110.

[9] Cf. Luc Deliens, Freddy Mortier, Johan Bilsen, Marc Cosyns, Robert Vander Stichele, Johan Vanoverloop and Koen Ingels, "End-of-life Decisions in Medical Practice in Flanders, Belgium: A Nationwide Survey," *Lancet*, Vol. 356 (November 25, 2000): 1806–1811.

[10] Personal communication on August 27, 2000.

morphine to ease the pain and the patient dies. However, in cases where the patients were competent and the doctor took their lives without asking them, this is worrisome. Some of the doctors' justifications for their actions with these patients were invalid as was the apparent unwillingness of some doctors to consult a colleague.[11] Indeed, according to the 1990 report, among the considerations by the physician in performing a life-terminating act without explicit request of the patient were the patient's low quality of life (31% of the cases), inability of the patient's relatives to cope with the situation (32%), and economic considerations (1%).[12]

Egbert Schroten, who has served on many medical committees, says that the findings are worrisome but that the discussion about their seriousness is exaggerated. Such medical behavior always existed before the invocation of the euthanasia policy, and a good doctor needs to help patients who are in great pain. Suffering of body and spirit is the most important consideration, and here we speak of incompetent patients in the very last days of their lives, who suffer miserably.

Ruud ter Meulen indicates that the issue worries him. He would like to have in place an explicit policy on termination of life, with no room for interpretation. The policy should insist on the explicit request of the patient and on strict medical criteria. There is also a need to define the concept of suffering. Ter Meulen expresses a critical view of the existing practice of passive euthanasia and the use of morphine to shorten life. He argues that it is not clear on which criteria decisions for passive euthanasia are based and what role the patient has in the decision-making process.

As expected, the three most critical voices in the present survey did not justify the existing situation. Both Chris Rutenfrans and Henk Jochemsen expressed concern about the lack of control mechanisms and the freedom that physicians have to decide the fates of their patients without consulting them. Jochemsen acknowledged that some were probably cases of double effect, but in others patients were competent and were still not apprised of the situation. This is not morally justifiable. The major consideration for physicians must be the suffering of the patient. Autonomy gives an additional justification, but does not constitute the major reason. While recognizing that the termination of life took place in the last stage of the disease, Jochemsen emphasized that there was no preference expressed by patients in a significant number of cases.[13]

Most outspoken was Frank Koerselman, who contends that junior doctors readily make Do Not Resuscitate (DNR) orders without much thought, especially when patients are old. Young doctors often evaluate a patient's quality of life without even knowing the patient, and many of them do not find compelling reasons for working to save a 90-year-old patient.

11 Paul J. van der Maas, Gerrit van der Wal, Ilinka Haverkate *et al.*, "Euthanasia, Physician-Assisted Suicide, and Other Medical Practices Involving the End of Life in the Netherlands, 1990–1995," *op. cit.*, p. 1704, Table 4.

12 P.J. van der Maas, J.J.M. van Delden, and L. Pijnenborg, *Euthanasia and other Medical Decisions Concerning the End of Life*, *op. cit.*, p. 64.

13 According to the 1990 report, the patient had given some indication about terminating life in 28% of cases. Interestingly, this consideration is mentioned by physicians in only 17% of cases. Apparently, it was not so much the wish of the patient but the circumstances that made the physician appreciate the patient's wish. Cf. P.J. van der Maas, J.J.M. van Delden, and L. Pijnenborg, *Euthanasia and other Medical Decisions Concerning the End of Life*, *op. cit.*, p. 64.

Koerselman testifies that he has seen many cases in which DNR orders were taken by phone or given by a junior physician without consulting a senior colleague.[14]

FEARS OF ELDERLY PATIENTS

The next question was: "Some Dutch studies appear to indicate that some elderly people fear their lives will be ended without their consent[15] and that, in fact, families in the Netherlands request euthanasia more often than the patient.[16] Is this true?"

A study in thirty nursing homes showed that when medical indications for hospitalization of elderly patients arose, nursing home physicians decided not to transport the patient to the hospital in 12% of cases, particularly when there was a life-threatening emergency. In a considerable number of cases, the decisions were made without consulting the patients or their families.[17] In a study done in Dutch hospitals, doctors and nurses reported that more requests for euthanasia came from families than from patients. The family, the doctors, and the nurses often pressured the patient to request euthanasia.[18]

Herbert Cohen said in an interview with John Keown that he would be put in a very difficult position if a patient told him that he really felt he was a nuisance to his relatives because they wanted to enjoy his estate. Asked whether he would rule out euthanasia in such a case, Cohen replied that in the end he wouldn't because "that kind of influence – these children wanting the money now – is the same kind of power from the past that… shaped us all." Cohen maintained that the same thing goes for religion, education, family of origin, "all kinds of influences from the past that we can't put aside."[19] It is unclear how this view could be seen as an acceptable interpretation of the Guidelines that speak of free and voluntary request of the patient as well as of unbearable suffering.

A minority of the interviewees dismissed the question as "ridiculous." Dupuis, van Dantzig and Keizer expressed the most supportive voices for the existing policy and practice of euthanasia.[20] Heleen Dupuis explained that when the patient is incompetent, the only available option for physicians is to cease treatment, not to perform euthanasia. She maintained that most people are afraid of having no option at the end of life and that in any event, old and demented patients receive treatment. There is no room for fear, and it is ridiculous to be afraid. In turn, van Dantzig dismissed the question as propaganda of the anti-

14 C.I. Dessaur and C.J.C. Rutenfrans wrote that a young doctor expressed objections to the use of pacemakers in people older than seventy-five years and declared that society may not be burdened with the duty of keeping old people alive. Cf. "The Present Day Practice of Euthanasia," *Issues in Law and Medicine*, Vol. 3, No. 4 (Spring 1988), at 402.

15 J.H. Segers, "Elderly Persons on the Subject of Euthanasia," *Issues in Law & Medicine*, Vol. 3 (November 4, 1988): 407–424; R. Fenigsen, "A Case Against Dutch Euthanasia," *Hastings Center Report*, Vol. 19 (Supp.) (1989): 24–26. Fenigsen argues (p. 24) that spouses have coerced their husbands or wives to undergo "voluntary" euthanasia.

16 R. Fenigsen, "A Case Against Dutch Euthanasia," *Ibid*, 22–30.

17 Richard Fenigsen, "Physician-Assisted Death in the Netherlands: Impact on Long-Term Care," *Issues in Law & Medicine*, Vol. 11, No. 3 (1995): 293–294.

18 Herbert Hendin, *Seduced by Death, op. cit.*, p. 93.

19 John Keown, "The Law and Practice of Euthanasia in the Netherlands," *The Law Quarterly Review*, Vol. 108 (January 1992), p. 63; *idem*, "Euthanasia in the Netherlands: Sliding Down the Slippery Slope?," *Notre Dame J. of Law, Ethics and Public Policy*, Vol. 9 (1995), p. 412.

20 In his comments on the first draft of this study, van Dantzig wrote succinctly: "I do not wholeheartedly support the existing system." Personal communication on July 6, 2000.

euthanasia lobby. Cases that include bad relations in a family should not prevent euthanasia in the many other cases where the motivation is sincere. However, the patient needs to express a will to die. If the patient does not express such a will and does not suffer, then it is not an appropriate case for euthanasia. If the patient had a living will, then the living will should be respected.

Bert Keizer said that it is not true that families request euthanasia more often than the patient. At least this is not his experience. Relatives may ask to shorten the dying process but doctors care about their patients and would never cut life short simply because relatives find it difficult to watch a dying patient. Keizer added that these requests take place at the end of life, during the last 2 or 3 days of life. As for the alleged fears of elderly people that their lives will be ended without their consent, Keizer regarded this as "crap, evil crap." He says that he is not going to answer such a question. He never sensed such fears among his patients.

Likewise, George Beusmans and Gerrit Kimsma who also practice euthanasia testified that in their medical practice they always provide information to the patient as well as to the family. It is irrelevant if the family wants to take a certain course of action when the patient does not share the same preference. It is only the patient's request that counts. When they practice euthanasia, it is done in the open with the consent of the patient, who must voice repeated requests and sign a written document. Kimsma asserts that his patients' families have always been satisfied with the euthanasia procedure. He has never had a family member object to euthanasia, and he has sometimes organized family meetings to resolve differences of opinion. At the same time, Kimsma declares that euthanasia is an individual matter of the patient and that he always resists family pressures for euthanasia when the patient does not want it.

Van der Wal also dismissed the question, saying that Segers who wrote about fears of elderly patients was speculating, and that the articles I cited are "nonsense" and "scientifically unsound." In one of the articles he co-authored, van der Wal wrote that Dutch studies do not provide any evidence for the elderly being in danger of becoming "victims" of euthanasia or assisted suicide.[21] Furthermore, van der Wal did not know of any study on the role of families (according to H.J.J. Leenen and Govert den Hartogh, no research has been done on the role of the family), and he and his research team were exploring the issue of conducting such research at the time of the interview. Van der Wal knows of only one opinion poll designed to investigate the issue, which, in his opinion, was conducted by a biased fundamentalist organization and was scientifically unsound. He further notes that euthanasia and assisted suicide among the elderly are rare. The age of most patients requesting euthanasia varies between 55 and 75.[22]

Having said that, although van der Wal dismisses the issue as "nonsense," he acknowledges that sometimes families find the suffering of their loved one unbearable and that there has been incidental anecdotal evidence of family pressure being brought to bear in

21 Martien T. Muller, Gerrit K. Kimsma and Gerrit van der Wal, "Euthanasia and Assisted Suicide: Facts, Figures and Fancies with Special Regard to Old Age," *Drugs & Aging*, Vol. 13, No. 3 (September 1998), p. 185.

22 Cf. P.J. van der Maas, J.J.M. van Delden, and L. Pijnenborg, *Euthanasia and other Medical Decisions Concerning the End of Life*, *op. cit.*, pp. 44, 137. In both tables, the first age category is 0–49; the second 50–64. The same age categories are used in Table 3 of the 1995 study. Cf. Paul J. van der Maas, Gerrit van der Wal, Ilinka Haverkate *et al.*, "Euthanasia, Physician-Assisted Suicide, and Other Medical Practices Involving the End of Life in the Netherlands, 1990–1995," *op. cit.*, p. 1703.

the termination of a patient's life. That is to say that family pressures do exist, but doctors testify that they are able to resist those pressures. Van der Wal emphasizes that, unlike other countries, there are no financial incentives for requesting euthanasia in the Netherlands because the costs of maintaining patients are not borne by their families.

H.J.J. Leenen says that he has never heard of patients who are afraid of euthanasia. On the contrary, one hears more about patients who fear that they will ask for euthanasia when their time comes, but their request will not be honored. J.K. Gevers and Govert den Hartogh do not think that families request euthanasia any more than patients do, or that physicians act on the wish of the family more than on the wish of the patient. In his comments on the first draft of this study, den Hartogh wrote that in all of the cases reported, the doctor acted on the request of the patient, and that in almost all instances the family supported the request of the patient. There is hardly any evidence of undue influence of the wishes of the family on the decision of the doctor, either from the reported cases or from other sources.

On the other hand, some physicians and consultants are insufficiently aware of the possibility of subtle forms of psychological pressure exerted on patients by their families. They routinely allow the family to be present at all discussions with the patient. This is one of the ways in which the SCEN-project (a network of consultants with whom physicians can consult on euthanasia cases) may lead to substantial improvement, as SCEN-doctors are trained to be alert to this possibility.[23]

Some interviewees argued that even if such fears on the part of patients exist, they are unfounded. Opinion polls show that the public thinks it is unacceptable for the family to request euthanasia because they cannot cope with the patient's condition. Families may request euthanasia for the patient because they might become exhausted by dealing with the patient's suffering and illness, but such a request would not be granted because testimony of a relative does not constitute sufficient grounds for euthanasia. Schroten, Wijsbek and van Leeuwen have heard that elderly patients fear their lives might be terminated prematurely without their consent. Schroten and Wijsbek believe that these fears often arise from religious convictions and are totally unrealistic. The three ethicists noted that euthanasia is rarely conducted in nursing homes. Indeed, studies show that Dutch nursing home physicians together receive an average of 300 requests for euthanasia and assisted suicide annually, of which they comply with only 25.[24] In comparison to general hospitals, nursing homes have a less permissive policy on euthanasia.[25] Bert Keizer writes that most of the patients in nursing homes belong to the generation that still associates "euthanasia" with "mass murder."[26]

Van Leeuwen maintains that there might be cases of inadequate care and, as a result, patients may die earlier than expected. Van Leeuwen further testifies that he once heard a

23 Personal communication of den Hartogh on August 27, 2000.

24 Martien Tom Muller, *Death on Request* (Amsterdam, Vrije Universiteit Thesis, 1996), p. 18. See also Table 3 in Paul J. van der Maas, Gerrit van der Wal, Ilinka Haverkate *et al.*, "Euthanasia, Physician-Assisted Suicide, and Other Medical Practices Involving the End of Life in the Netherlands, 1990–1995," *op. cit.*, p. 1703, and Table 2 in Gerrit van der Wal and P.J. van der Maas, "Empirical Research on Euthanasia and Other Medical End-of-Life Decisions and the Euthanasia Notification Procedure," in David C. Thomasma *et al.* (eds.), *Asking to Die, op. cit.*, p. 172.

25 Dick J. Hessing, John R. Blad and Roel Pieterman, "Practical Reasons and Reasonable Practice: The Case of Euthanasia in the Netherlands," *J. of Social Issues,* Vol. 52, No. 2 (1996), p. 155.

26 Bert Keizer, *Dancing with Mister D* (London: Black Swan, 1997), p. 258. Keizer also testifies that only a small percentage of his patients *really* want to die (emphasis mine, RCA). *Ibid.*, p. 268.

physician describing how he and his colleagues could have improved a patient's situation, but did not do so because the patient had no family to care for him were his condition to improve, thus leaving him alone in a difficult situation.[27] In addition, when some families request that the elderly die peacefully, they are indirectly exerting their influence on the decision-making.

Parenthetically, let me refer to Bert Keizer's exchange with a nephew of one of his patients, who assured him that "this is not at all what he [the patient] wanted, ending his days in a place like this." Keizer, who practices medicine in a nursing home in Amsterdam, writes: "it always annoys me, that tone of voice in which people say, Uncle would never have wanted this. What they mean is, he's not as stupid as those other 56,000 Dutch people staying in nursing homes who have 'let things go too far'."[28]

Van Delden, one of the authors of the 1990 comprehensive study, based his answer on this study. The study shows that relatives made an explicit request to hasten the death of the patient in 14% of the cases, and others (physician, nurse or someone else who is not a relative) made the request in 1% of cases.[29]

Van der Arend tells of arguments he witnessed between relatives over financial affairs, even about an inheritance while the patient was still alive. Therefore, families should be engaged in the decision-making, but should not be given the last word. All parties should be involved in the euthanasia decision: the patient, the physicians and nurses, and the patient's family. If members of the family are not involved, they might end up going to court. Van der Arend added that patients usually trust physicians not to terminate their lives without their consent. When the patient is rendered incompetent and has not prepared a living will, the family assumes a larger role in the decision-making process, but still does not have the last word.

Rob Houtepen and Ron Berghmans argue that families do exert pressure for, but also against, euthanasia. Most physicians are aware of the pressure on them to perform euthanasia, and they resist it. The common procedure is to inform and to speak with family members. They estimate that families are consulted in the majority of euthanasia cases.[30] In turn, John Griffiths identified the issue of family pressure as one of the problems that needs to be addressed by effective regulation. Families that can no longer cope with the patient's situation might exert pressure on doctors and influence them to opt for euthanasia rather than to provide adequate pain relief. Griffiths maintained in a later communication that this

27 In his comments, van Leeuwen asked me to emphasize that he heard this once, and only once. Personal communication on August 30, 2000.

28 Bert Keizer, *Dancing with Mister D*, pp. 301–302. See also pp. 306, 312.

29 P.J. van der Maas, J.J.M. van Delden, and L. Pijnenborg, *Euthanasia and other Medical Decisions Concerning the End of Life*, *op. cit.*, p. 134.

30 According to the 1990 study, GPs consulted with patients' relatives in 97% of the cases, and specialists consulted with relatives in 85% of cases. Cf. P.J. van der Maas, J.J.M. van Delden, and L. Pijnenborg, *Euthanasia and other Medical Decisions Concerning the End of Life*, *op. cit.*, p. 47. According to the 1995 death-certificate study, "relatives or others" were consulted only in 70% of euthanasia and assisted-suicide cases. Paul J. van der Maas, Gerrit van der Wal, Ilinka Haverkate *et al.*, "Euthanasia, Physician-Assisted Suicide, and Other Medical Practices Involving the End of Life in the Netherlands, 1990–1995," *op. cit.*, p. 1704. Interestingly, another study shows that patient's relatives were "contacted" in more than 90% of the euthanasia and assisted-suicide cases. Gerrit van der Wal, Paul J. van der Maas, Jacqueline M. Bosma *et al.*, "Evaluation of the Notification Procedure for Physician-Assisted Death in the Netherlands," *New Eng. J. of Med.*, Vol. 335, No. 22 (November 28, 1996), p. 1709.

problem is in no way limited to euthanasia. His impression is that it is far more important quantitatively in connection with abstention decisions.

Ruud ter Meulen, Henk Jochemsen and Frank Koerselman voiced the most critical voices on this matter. Ter Meulen is worried about the position of elderly people, given that there are waiting lists for nursing homes and for home care. The Netherlands is facing a shortage of resources, cuts in the nursing staff and a larger number of the elderly who wish to be in nursing homes. We can expect that the quality of care for the elderly will deteriorate. Euthanasia may be a promising alternative as a solution to a pressing problem, and ter Meulen adds that this is especially true with respect to elderly patients who do not want to be a burden on their families. This consideration may play a role in their request for euthanasia. However, in his comments on the first draft of this paper, ter Meulen added that there is no evidence of this issue playing a role in the practice of euthanasia at this point.

Henk Jochemsen contends that it is often the case of the family, rather than the patient, asking for euthanasia when the patient has become a burden. Hence, Jochemsen thinks that some elderly patients indeed have good reason for this fear. He adds that patients sometimes also fear that treatment will continue beyond the necessary point. Jochemsen also objects to providing unnecessary treatment.

Koerselman expressed the most critical view, agreeing with the studies of Segers and Fenigsen, and testifying about his own experience. He argued that such fears on the part of the elderly do exist. In his view, families have informal influence on doctors, and it is impossible to ascertain that families who find it difficult to cope with the suffering of their loved ones do not influence the decision-making process. Koerselman told the story of an 85-year-old patient with pneumonia and depression, both of which are treatable conditions. His family did not wish to treat him, and the GP agreed with the family that there was no point in treatment. Koerselman was invited to consult on the patient's depression and said that if he had not taken an active role in this case, they would have taken the easy way out: "He would die, all would say that I acted humanely and would receive a nice bottle of wine from the family who physically wanted to prevent me from treating him." Koerselman asked the patient's GP whether he had spoken with the patient alone. The response was negative. Koerselman had to order security guards to remove the family from the room so that he could speak privately with the patient. The patient said at first that there was no reason to continue treatment and that he did not want to become a burden on his family. Koerselman explained to him that he suffered from pneumonia and depression, two perfectly treatable conditions, and that he might as well try to be a healthy 85-year-old man. The patient then agreed to treatment and was eventually discharged from the hospital in very good condition, both mentally and physically.

CHAPTER 7

THE REMMELINK CONTENTION AND THE BRITISH CRITICISM

THE REMMELINK CONTENTION

The next question was: The Remmelink Commission held that actively ending life when the vital functions have started failing is indisputably normal medical practice.[1] Is this correct? What is your opinion?

When I first read this statement, I was puzzled. It is unclear what "the vital functions have started failing" exactly means. What vital functions? What does "started failing" signify? Moreover, is this really the common practice in the Netherlands? The assertion is unqualified. The consent of the patient does not appear in it. To say that it is "indisputably normal medical practice" seemed to be quite dangerous.

Interestingly, some interviewees denied that the Report actually said this. Most of them disagreed with the unqualified statement. Two interviewees agreed with the statement and two others understood it to refer to double effect, which is an acceptable doctrine in the Netherlands as well as in other countries.

Evert van Leeuwen and Heleen Dupuis disagreed with the Remmelink statement and added that they did not think the Commission had actually expressed such a vague statement. "Failing of vital functions" is not a common phrase, and it does not convey a clear meaning. They questioned the statement's rationale: If vital functions are failing, that means the patient is dying. If someone is dying, why is there a need to kill him? Heleen Dupuis wondered: What does "vital functions" mean? If the statement refers to patients in a coma, the practice is to stop treatment after several months. But this rule as stated here "is nonsense." The doctor must know more in order to terminate life. Three interviewees, Schroten, Koerselman and van Delden, said that if the vital organs are irreversibly failing, then the patient is dying. Switching ventilators is a normal practice, but not injections designed to kill. Medical treatment should stop under such circumstances, but this does not mean actively ending life. In any event, lethal injections are not normal treatment, and euthanasia is not a normal practice. It is an exceptional treatment used in cases of exceptional suffering of the patients concerned.

Van Delden was familiar with the statement, which accompanied the 1990 study report that he co-authored with van der Maas and Pijnenborg, and expressed disagreement with it. He emphasized that there were two documents: one of the Commission and one of the van der Maas research group, in which he participated. This contention was made in the Commission's Report, and this part of the Report was not accepted by the Dutch parliament. Van Delden explained that the contention was made by the Commission to justify what was

1 Henk A.M.J. ten Have, "Euthanasia: The Dutch Experience," *Annals de la Real Academia Nacional de Medicina,* Tomo CXII (Madrid, 1995), p. 429.

going on, but that this was not something he would condone. Likewise, van der Wal was familiar with the statement and thought that the Commission erred in including it. In his view, this went a step too far, and he did not endorse this reasoning. He asserted that euthanasia is *not* a normal practice and should be avoided as much as possible. As such, we pass a boundary when we say that it is normal and accept ending life without an explicit request of the patient.

Arie van der Arend also did not agree with the Remmelink statement, saying that physicians in general do not want to terminate life. They want to do whatever they can to save the lives of their patients. Preserving life is the normal medical practice, and he doubted that the Remmelink Commission actually made the above statement. In his view, the normal practice, if any at all, with respect to hopeless situations is to withhold treatment, not to actively end the life of the patient. He testified from personal experience in the neonatal department that withholding treatment occurs, but not active euthanasia. Van der Arend is convinced that doctors in the Netherlands do not accept or follow the Remmelink statement.

Bert Thijs remarks that the entire function of the Intensive Care Unit (ICU) is to try to save patients whose vital functions have failed, whose breathing is difficult, and whose blood pressure has dropped considerably. The normal practice in all ICUs, including the one that he directs, is to try to save the lives of these patients.

Henk Jochemsen, H.J.J. Leenen, Bert Keizer and Govert den Hartogh expressed strong disagreement with the Remmelink statement. Jochemsen explained that the Remmelink statement concerned the 900 and 1000 patients who had not given their consent. He argued that actively shortening life is not normal medical practice and that the government and the courts do not conceive of this as normal medical practice. After all, if such a practice is considered normal, then why report? Why control? It does not make sense.

Leenen insisted time and again that euthanasia is not normal medical practice, and that the Remmelink's view is absolutely unacceptable in the Netherlands. The Guidelines speak of autonomous decision-making, whereas the Remmelink statement does not refer to autonomy. In his later comments, Leenen asserted that the Remmelink contention had no relevance in the debate. Hence, "why give it so much accent?"[2] Den Hartogh explained that when dealing with patients whose vital functions are failing, doctors should cease treatment but continue with palliative care. They should not actively end life. He maintained that the Remmelink statement is contrary to what is accepted today in the Netherlands and that euthanasia should remain an exceptional medical practice conducted in cases of unbearable suffering.[3]

On the other hand, Henri Wijsbek and A. van Dantzig agreed with the Remmelink statement, arguing that it is senseless to continue treatment and medications when vital functions start failing. Although there is no consent on the part of the patient, life beyond repair is senseless and euthanasia is permitted under these conditions. Van Dantzig explained that "failing vital functions" means keeping a person alive by external means, such as respirators and heart stimulation. The welfare of the patient is the main concern, and in such

2 Letter dated July 25, 2000.

3 Similarly, Rob Houtepen, Ruud ter Meulen and Ron Berghmans rejected the contention, saying that no form of active termination of life is normal medical procedure. Rather, euthanasia is an exceptional practice and should remain so.

severe circumstances the patient should be allowed to die. In turn, John Griffiths explains that "normal medical practice" is a legal term referring to the behavior (otherwise illegal) that doctors can perform by virtue of the authority to practice medicine. The statement refers to people who would have died within *hours,* who were suffering an irreversible failure of all functions, and whose doctors shortened the process of death. Upon my expressed pity that Remmelink did not state all of this explicitly, Griffiths responded that Remmelink thought the point was obvious and, therefore, did not explain.[4]

J.K. (Sjef) Gevers also understands this statement to mean providing extra morphine to ease suffering during the final hours of the patient's life. In his mind, it is a grey area that needs to be discussed, not a matter only for doctors to decide. Gevers did maintain that it would have been preferable to elaborate on this statement, but unfortunately Remmelink did not do so.

I found it most troublesome that the person who was nominated by the Dutch government to study the practice of euthanasia had made the above ambiguous statement, disregarding one of the basic components required for the euthanasia practice: the will of the patient. Many of the interviewees also found this statement troublesome for similar reasons. Critics of the Dutch practice might possibly note that the unqualified statement reflects a certain dangerous culture that is unhealthy for patients, culture that tends to forego life too easily.

THE MEMORANDUM OF THE BRITISH MEDICAL ASSOCIATION

One of the most thorough investigations of euthanasia and physician-assisted suicide (PAS) is that of the British House of Lords Select Committee on Medical Ethics. In their lengthy report, the Select Committee interviewed many interested individuals and parties and detailed the main arguments pro and con allowing mercy killings.

Among the interested parties was the British Medical Association (BMA) that voiced its strong opposition to allowing euthanasia or PAS in Britain. A key argument in their critique of those two practices was the apparent failure of the Dutch policy on mercy killings. In its memorandum before the House of Lords, the BMA held that in regard to Holland, "all seem to agree that the so-called rules of careful conduct (official Guidelines for euthanasia) are disregarded in some cases. Breaches of rules range from the practice of involuntary euthanasia to failure to consult another practitioner before carrying out euthanasia and to certifying the cause of death as natural."[5] I asked my interviewees: Do you agree?

This statement is straightforward, and anyone who is familiar with the Dutch policy and practice should concede that it is true. I wanted to see whether the protective mechanisms of the Dutch policy employed by some of the interviewees might bring them to deny this assertion, and on what grounds.

4 In his comments on the first draft of this essay, Griffiths referred me to p. 132 of his book, *Euthanasia & Law in the Netherlands*, in which he relates to the Remmelink contention about patients whose bodily functions are "successively and irreversibly failing." Griffiths thinks that understanding the contention in these terms makes a big difference, claiming that "you will find that doctors know perfectly well what this means, and furthermore that it is standard practice in a large number of countries and has been for many years (although only in the Netherlands is anyone prepared to talk about it)." Personal communication on July 10, 2000.

5 House of Lords, *Select Committee on Medical Ethics*, session 1993–94, Vol. II, Minutes of Oral Evidence (London: HMSO, 1994), at 33.

Many interviewees conceded that this assertion is, indeed, correct.[6] Two interviewees disregarded the issue and three others said that the assertion is both true and untrue. Van der Arend and Berghmans agreed with the assertion and added that what is needed is more education on the rules of careful conduct. But, they said that we need to recognize that there will always be physicians who will not follow the Guidelines, and they should be prosecuted. They maintained that these are marginal cases and that, essentially, the practice is no different from what is happening in other countries in a more secretive way. In a similar fashion, their colleague Rob Houtepen agreed with the British statement but added that a fair appraisal of the Dutch euthanasia practice required a comparison with other countries. Bert Keizer also conceded that the British statement is correct. Indeed, only fifty per cent of euthanasia cases are reported in the Netherlands. But, Keizer added, we know nothing about what is going on in the United Kingdom; there the reporting rate is zero. In the Netherlands, at least, "we know part of the picture and we are working to improve the situation." Van Leeuwen said that he does not think we can blame physicians for intentional killing. More simply, not all physicians are aware of the need to consult. Hence, more education and explanation of the procedures are required. Egbert Schroten asserted that in the 1980s and the beginning of the 1990s, not all doctors knew the exact wording of the Guidelines. They did not know that they needed to consult a colleague. Now, in 1999, things are clearer and most of the doctors do consult a colleague.

Similarly, van Delden acknowledged that the British statement is factually true. He added that it is difficult to move from justified individual cases to policy making. There will always be people who abuse their power. Van Delden explained that a pertinent distinction is between *content* Guidelines and *procedural* Guidelines. Content Guidelines refer to such aspects as the condition of the patient and the expression of a reiterated voluntary request, whereas the procedural Guidelines refer to the notification procedure and consultancy. Van Delden holds that many un-notified cases meet the content Guidelines, but not the procedural Guidelines. This happens because doctors fear the hassle involved in reporting, respect the privacy of their patients, and do not wish to be scrutinized.

Ruud ter Meulen indicated that the British criticism is correct and that the current situation is the result of unclear Guidelines (like van Delden, he wonders what does "unbearable suffering" mean) and lack of control, which bring some doctors to disregard the criteria for careful conduct. Frank Koerselman added that what is most troublesome is not the fact that the Guidelines are broken, but that they keep changing and becoming more receptive to euthanasia. It is the climate that worries him. Similarly, Henk Jochemsen is worried about the missionary vigor employed by the establishment to defend the policy and practice of euthanasia.

In his comments on the first draft of this study, Arko Oderwald wrote that the cases of involuntary euthanasia are worrisome and that further research is necessary. In his view, the most troublesome fact is the tendency to allow doctors to act on their own without somebody, in the name of society, watching over their shoulder. If doctors are as honorable as they say, they have nothing to hide and they should understand this issue as a social issue, not as a purely medical or personal issue.[7]

6 John Griffiths, Evert van Leeuwen, Henri Wijsbek, Ruud ter Meulen, Bert Keizer, Arie van der Arend, Henk Jochemsen, Chris Rutenfrans, Frank Koerselman, Egbert Schroten and Rob Houtepen.

7 Personal communication on August 28, 2000.

As said, two interviewees did not pay much notice to the issue and three others said that the British statement is both true and untrue. Van Dantzig and Dupuis remained most protective of the Dutch policy and practice of euthanasia. Van Dantzig answered that he did not "know about the British statement." He didn't wish to relate to the *content* of the statement.[8] Heleen Dupuis, who is professor of medical ethics, disagreed with the BMA's statement, saying that she could not imagine doctors who fail to consult a colleague when euthanasia is concerned. She acknowledged that sometimes not all the Guidelines are satisfied but insisted that the same happened all over the world. That some doctors do not observe all the Guidelines does not mean that the Guidelines are wrong. What is needed is to punish the sloppy doctors.

This line of questioning provoked Dupuis to make the following statement: "Doctors try to save life, not to kill patients. Why should doctors kill their patients? What would be the motive? After all, the doctor knows he would go to jail if he does. It is ridiculous to assume this." She maintained that the phenomenon of "angels of death" never happens in Holland.[9] It happens in countries that do not discuss euthanasia in the open, "but not here where everything is in the open," discussed and under scrutiny.[10]

Henk Leenen, Govert den Hartogh and Gerrit van der Wal said that the British statement is both true and untrue. Leenen only agreed with the British contention regarding the lack of reporting.[11] Den Hartogh argued that there is no proof of involuntary euthanasia, but it is true that there have been cases in which physicians failed to report and to consult. In his comments on the first draft of the study, den Hartogh added that he is sure that no "practice" of involuntary euthanasia (i.e. "euthanasia" contrary to the wishes of the patient, doctors getting rid of unwanted patients) exists because there is not the slightest trace of evidence of it, and it would be impossible for such a practice to exist without a trace. However, non-voluntary euthanasia does, of course, occur as documented by the two van der Maas reports.[12] Van der Wal explained that in 1995, the consultation rate was 63% and now it is

8 In his comments, van Dantzig wrote: "I hope that you will be so kind as to remove all linkage between Prof. Dupuis and myself. She is a friend of mine, and I hold her in the highest regard, but on some things we are of different opinions." Personal communication on July 6, 2000.

9 In "A Case Against Dutch Euthanasia," *op. cit.*, pp. 24–25, Fenigsen argued that in 1987, a series of killings of comatose patients took place at the department of neurosurgery at the Free University Hospital in Amsterdam. Four nurses were responsible for these serial killings. Furthermore, a doctor was apprehended in The Hague under suspicion of having killed twenty inhabitants of the De Terp old people's home without their consent or knowledge. He pleaded guilty to five, was accused of four, and convicted of three killings. Witnesses testified that some of the victims were not ill but only senile and querulous, and that the doctor was impatient with elderly people, reluctant to treat them, frequently absent, and left many decisions to the male head nurse. Hendin writes on angels of death, a team of travelling physicians that provided euthanasia to patients when family doctors were unwilling to do so. See Herbert Hendin, *Seduced by Death, op. cit.*, pp. 110–113.

10 Interestingly, Koerselman comments in this regard that the prime obligation of doctors all over the world is to fight for life, to save life. This is not the case in the Netherlands. Here the prime consideration is to relieve suffering. Arko Oderwald says that doctors do not like to do nothing; they like to act in order to change the situation. If there is nothing to do medically but to wait for death and to care for the patient, they might be prompted to perform the last act that is still available: euthanasia.

11 This was reiterated in Leenen's letter dated February 2, 2001, commenting on a draft of this study.

12 Personal communication on August 27, 2000.

higher.[13] Consultancy has to include experienced doctors actually seeing the patient. I asked whether consultation is carried out over the phone, and van der Wal answered "I don't know. Possibly yes." This calm and calculated tone of the answer, which was repeated in many interviews, worried me. I shall discuss this issue in *chapter 9.* But let me first reflect on the question of whether physicians should suggest euthanasia to their patients.

[13] According to the physicians' interviews in the 1995 survey, physicians consulted with a colleague in 93% of the reported cases, but in only 18% of the unreported cases of euthanasia and assisted suicide. Gerrit van der Wal and P.J. van der Maas, "Empirical Research on Euthanasia and Other Medical End-of-Life Decisions and the Euthanasia Notification Procedure," in David C. Thomasma *et al.* (eds.), *Asking to Die, op., cit.,* Table 6, p. 176.

CHAPTER 8

SHOULD PHYSICIANS SUGGEST EUTHANASIA TO THEIR PATIENTS?

INTRODUCTION

Before asking about consultation, I asked the interviewees to focus on one issue that I consider as ethically problematic: Whether a physician should suggest euthanasia or assisted suicide to his or her patients. The KNMG report on euthanasia describes a situation in which the physician has the impression that the patient would like to start a conversation about the end of his life and his wishes concerning the end of his life, but hesitates to start this conversation. If this is the case, then the physician might choose to start this conversation. However, this has to be done with the utmost precaution in order to avoid making the patient feel pushed in a certain direction.[1] The 1990 study shows that 36% of specialists, 24% of home physicians, and 65% (!) of general practitioners believed that there can be situations in which the physician should raise euthanasia as a possibility with the patient.[2] The 1990 prospective study shows that the initiative for discussion about the action to be performed at the end of life came from the patient in only about half of the cases.[3] Van der Maas and van der Wal estimated that of all cases of euthanasia, PAS and the ending of life without the patient's explicit request, the physician initiated the discussion in 21 percent.[4] Another study holds that 54 percent of physicians believe that in certain situations it is the physician's professional duty to raise euthanasia as an option with the patient.[5] Neither the physicians nor the study's investigators seem to acknowledge to what extent the voluntariness of the process may be compromised by such a suggestion.[6]

Pieter Admiraal revealed that "contrary to most doctors," who will not discuss euthanasia before the patient requests it, he discusses this option with the patient, "for it can be of great value and great comfort for many." Admiraal further explains that not discussing euthanasia with a patient can take the medical team by surprise if a patient suddenly asks for it later. One might question whether this issue is of real concern. Admiraal argues that if one considers euthanasia permissible under some circumstances, then not to broach the

1 Inzake Euthanasie. Koninklijke Nederlandsche Maatschappij tot bevordering der Geneeskunst (1995), p. 15. ISBN 90-71994-10-4. See also "Discussienota van de Werkgroep Euthanasie," van de Koninklijke Maatschappij tot Bevordering der Geneeskunst (KNMG), *Medisch Contact*, Vol. 30 (1975): 7–16.

2 P.J. van der Maas, J.J.M. van Delden, and L. Pijnenborg, *Euthanasia and other Medical Decisions Concerning the End of Life*, Health Policy Monographs (Amsterdam: Elsevier, 1992), p. 102.

3 *Ibid.*, p. 156.

4 Paul J. van der Maas and Gerrit van der Wal, A letter to the Editor, *New Eng. J. of Med.*, Vol. 336, No. 19 (May 8, 1997), p. 1386.

5 Paul van der Maas and Linda L. Emanuel, "Factual Findings," in L.L. Emanuel (ed.), *Regulating How We Die* (Cambridge, Mass.: Harvard University Press, 1998), p. 168.

6 Herbert Hendin, Chris Rutenfrans and Zbigniew Zylicz, "Physician-Assisted Suicide and Euthanasia in the Netherlands," *JAMA*, Vol. 277, No. 21 (June 4, 1997), at 1721.

possibility with a patient is to deny that patient the full range of available options. Suggesting euthanasia to a patient is thus a measure of respect for the patient's autonomy.[7] Herbert Cohen likewise claims that raising the subject of euthanasia by the physician has an emancipating effect.[8]

My own inclination was to contest these arguments for the following reasons. The Dutch health care system is built around the general practitioner that has known his/her patients for many years. Based on this history of trust and confidence, the GP might feel comfortable raising the issue of euthanasia with the patient. This might have devastating implications—the doctor whom the patient has trusted for so long has nothing to offer but death. Knowing that the physician has given up might cause the patient to give up as well, to surrender his or her life. What if the patient wishes to continue living? Could the patient still trust a physician who offers to kill him? Could the patient trust that the GP would do everything possible to fight to maintain that patient's life? This might create a very uncomfortable situation for both physician and patient.

Furthermore, in the Netherlands, the patient has few options for changing the GP. Usually patients build long term relationships with their GPs, relationships that last decades during which trust and confidence are crystallized. Patients are hesitant to replace the familiar and trusted physician with someone else. Clearly, one does not seek more complications in the end-of-life fragile stage, and it would be difficult to build trust in a new physician. Moreover, in the Netherlands' cultural atmosphere, where patients generally do not want to become a burden on their families, an offer of euthanasia by their GP might be taken as a sign that they are living on borrowed time, which the GP does not find useful or of high quality. The GP's offer might lead not only to giving up the fight but also to increasing the patient's feelings of guilt for still being alive.[9]

For all these reasons, it was interesting to hear the thoughts of the interviewees about this issue, and whether they were at all aware of this ethical problem. Some interviewees did not share my concerns. For various reasons they thought physicians should offer euthanasia as an option. Two interviewees asserted that doctors don't propose euthanasia to their patients. Five interviewees objected to physician's initiative.

THE INTERVIEWEES' RESPONSES

John Griffiths' stated that there is no basis for Dutch doctors to propose euthanasia to their patients. He added that although euthanasia is not offered to patients, sometimes it would actually be better to propose it. Later in the interview, Griffiths acknowledged that suggesting euthanasia does happen, albeit infrequently and not on the level of constituting a social problem. In his comments on the first draft of this essay, Griffiths clarified:

7 Carlos F. Gomez, *Regulating Death* (New York: The Free Press, 1991), p. 109.

8 Herbert Hendin, *Seduced by Death* (New York: W.W. Norton, 1997), p. 52.

9 On the quality of life concept in medicine, see R. Cohen-Almagor and M. Shmueli, "Can Life Be Evaluated? The Jewish Halachic Approach vs. the Quality of Life Approach in Medical Ethics: A Critical View," *Theoretical Medicine and Bioethics*, Vol. 21, No. 2 (August 2000): 117–137. On the power of the word and certain terminology, see R. Cohen-Almagor, "Language and Reality at the End of Life," *J. Law, Medicine & Ethics*, Vol. 28, No. 3 (Fall 2000): 267–278, and see Andrea Nye (ed.), *Philosophy of Language* (Malden, Mass.: Blackwell, 1998).

> What I did say is that the suggestion by Hendin *et al.* to the effect that doctors steer their patient's decision making is based on a misreading of the relevant data (which refer only to 'first raising the question'). The latter seems to me good medical practice, in certain circumstances. It does not necessarily imply any sort of pressure or suggestion on the doctor's part. Whether the latter occurs, I do not know (I assume it does, doctors being human and therefore not always perfect). I do think an effective control system needs to guard against the risk.[10]

On the other hand, Arie van der Arend, a medical ethicist from Maastricht, did not think that these were rare incidents. He made a distinction between the physicians' initiation in offering euthanasia as an option, and the physicians' interpretation of patients' indications. While physicians should never offer euthanasia as an option, they should inquire about what their patients want.[11] Physicians may mention euthanasia only after the patient has indicated thinking about it and has discussed the issue while using other terms. In some cases, patients may be reluctant to raise the issue *per se*, and the physicians are required to label what the patients have indicated.[12]

Paul van der Maas who played a major role in the writing of the three major reports of 1990, 1995 and 2001, said that most physicians wish to include the euthanasia option in medicine. He thinks that it is important for the physician to be open to discussing all of the options with the patient, making it clear that the euthanasia option is available. If the patient does not initiate the discussion, then the doctor should do so. Van der Maas thinks that such a conversation is necessary when the physician does not know what the patient wants, just in case the patient loses consciousness. In a letter written with van der Wal in response to Hendin's critique, van der Maas explains that taking the initiative to create an opportunity for patients to discuss their wishes concerning the end of life is very different from Hendin's portrayal of "telling the patient that his or her life is not worth living."[13]

Two physicians who have been practicing euthanasia for many years, George Beusmans and Gerrit Kimsma, did not share my concerns about driving patients to opt death by raising the euthanasia option. Beusmans underestimated the ethical concerns, while Kimsma was aware of them but stressed the overriding principles that justify the physician's initiation of a discussion on euthanasia. Both of them believe it necessary for the physician to raise the issue to the patient because some patients may feel uncomfortable raising it themselves. Beusmans and Kimsma, in turn, feel comfortable raising the issue and consider it part of their role as doctors.

The two doctors elaborated on their conduct well beyond the framework of my question and gave interesting accounts that warrant a detailed description. Beusmans explained that most of his patients expressed appreciation for his initiation of the conversation. This is a very difficult issue for patients, who wish to be perceived as strong. They view the initiation

10 Personal communication on July 10, 2000.

11 Cf. Richard B. Balaban, "A Physician's Guide to Talking About End-of-Life Care," *Journal of General Internal Medicine*, Vol. 15, Issue 3 (March 2000): 195; A. Chopra, "Communicating Effectively at the End of Life," *J. Am Osteopath Assoc.*, Vol. 101, No. 10 (October 2001): 594–598; S. Z. Pantilat, A. J. Markowitz, "Perspectives on Care at the Close of Life. Initiating End-of-Life Discussions with Seriously Ill Patients," *JAMA*, Vol. 285, No. 22 (2001): 2906.

12 In a similar fashion, Ron Berghmans and Ruud ter Meulen, medical ethicists from Maastricht, voiced their objections to the doctor's initiation of discussion, but qualified their answers by saying that they could imagine exceptional cases to the general proscription of leaving the issue to be raised by the patient. The physicians should not raise the issue immediately, but they might mention it during the process approaching the end of life.

13 A letter to the Editor, *New Eng. J. of Med.*, Vol. 336, No. 19 (May 8, 1997): 1386.

of a discussion on euthanasia as a sign of weakness, indicating that the patient is unable to cope with the suffering. Beusmans articulated that in the first discussion on euthanasia, he does not say much. He offers it as an option and suggests that the patients consult with their families. This first discussion usually takes place when Beusmans estimates that the patient has two to three months to live. He likes to raise the issue then so as to enable the patients some time to crystallize their decision. The patients are more capable of thinking clearly without experiencing a lot of pain that might obscure critical reflection. Two weeks later, Beusmans discusses the options available to the patients: assisted suicide, euthanasia, or optimal treatment. The problem is not only pain. Many patients are fearful of dying and do not know what to expect. Physicians can handle the pain, but it is much more difficult to handle the mental aspects. Ultimately, the patients decide what they want.

Beusmans testifies that some 10 of his patients in the past decade have asked for his help in terminating their lives: nine asked for physician-assisted suicide and one for euthanasia. The one who asked for euthanasia was very sick, could not take any food and fluids, and asked for a lethal injection. Beusmans said that, on the whole, he does not like to give lethal injections.[14] Probably his reluctance to do so influenced his patients to choose physician-assisted suicide. Most Dutch physicians who perform PAS and euthanasia do not exhibit such reluctance, and consequently there are far more cases of euthanasia than cases of PAS.[15] Beusmans performs physician-assisted suicide when the patient's loved ones (usually 3–4 people) are present. All of these cases took place at the patients' homes with their families and/or friends present. Beusmans knew the patients and their families for years. All of them were cancer patients; all were competent and conscious.[16]

Gerrit Kimsma portrays a picture that is similar in some aspects and different in others. Similarly, all of his euthanasia patients had cancer. He knew them for a long time, most of them for more than 10 years. Most of his patients, "maybe all of them," were men. Unlike Beusmans, Kimsma says that in 12 years of performing euthanasia and PAS, only a handful of cases involved PAS.[17] Kimsma sees no substantive ethical difference between the two

14 Likewise, a family of K who was euthanized told me that their GP was relieved when K was able to take the lethal medication orally. The GP was willing to give the lethal injection if necessary, but was greatly comforted when the patient took the glass and drank the medication by himself. The de B. family, in turn, was reassured by the act that this was, indeed, what K wanted.

15 Herbert Hendin, *Seduced by Death, op. cit.,* p. 53. One study, which compared attitudes and practices concerning end-of-life decisions between physicians in Oregon and in the Netherlands, showed that an equal proportion of Dutch physicians considered euthanasia and PAS as ethically acceptable. Conversely, American physicians were consistently less likely to find euthanasia acceptable as compared with PAS. Cf. Dick L. Willems, Elisabeth R. Daniels, Gerrit van der Wal, P.J. van der Maas and E.L. Emanuel, "Attitudes and Practices Concerning the End of Life: A Comparison Between Physicians from the United States and from the Netherlands," *Arch Intern Med.,* Vol. 160 (2000): 63–68. See also R.J. Kohlwes, T.D. Koepsell, L.A. Rhodes, and R.A. Pearlman, "Physicians' Responses to Patients' Requests for Physician-assisted Suicide," *Archives of Internal Medicine*, Vol. 161 (March 12, 2001): 657–663; R. Cohen-Almagor and Monica G. Hartman, "The Oregon Death with Dignity Act: Review and Proposals for Improvement," *Journal of Legislation*, Vol. 27, No. 2 (2001): 269–298.

16 Beusmans added that if the patient "agrees with me to do it, and then falls into [a state of] unconsciousness, maybe I would do it." He continued to tell the story of one patient who was in a coma and very unquiet. "The family asked for something. I gave her morphine and after one day she died. It was a normal dose and I did not expect her to die so soon."

17 I asked Dr. Kimsma how many euthanasia cases he actually performed, and his answer was that the number is not important. It is the quality, not the quantity that is important. He said that he has been performing euthanasia

practices: The effect and motive are exactly the same, and only the method is different. At the same time, he acknowledges that there is a moral difference for him as a physician between injecting the lethal medication and providing the patient with the drugs. He asserts that physicians choose the needle more than the cup (oral medication) because they do not doubt the need for euthanasia and because the patients ask for it. Kimsma says that most of his own patients asked to die by the needle as soon as possible. As with Beusmans, the patients' choices probably reflect the physician's preferences.

Kimsma thinks that doctors need to promise their patients that they will not abandon them. Patients should not be left alone suffering. When a potentially terminal disease is found, the doctor should discuss the issue of euthanasia with the patient. Accordingly, Kimsma holds conversations about euthanasia with all of his 'terminal' patients. His opening remark to patients who have been informed of their terminal illness is: "Now you have heard this diagnosis, it may be that you have expectations or ideas about the end of life. If you wish to discuss these, I will be available for you, now or in some future time, whenever you feel the need."[18]

Kimsma also advises his patients to join the Dutch Voluntary Euthanasia Society.[19] He maintains that in the Netherlands "we have physicians who never talk about end-of-life issues" because they are afraid that the patient will start a discussion on euthanasia. Doctors and patients alike should be sensitive and open in discussing end-of-life issues. Physicians should discuss with their patients what the expectations are and which options they would like to consider. Sometimes patients do not, or will not, talk about euthanasia. In Kimsma's view, in 80% of the cases, physicians wait until the patient starts talking about euthanasia. In 10% of the cases, physicians are the ones who initiate the conversation, and in another 10% of the cases, it is the families that initiate discussion.

On too many occasions, because everyone is reluctant to initiate the conversation, it is postponed and then the physicians need to act under pressure. Kimsma testifies that because physicians wait too long, in 13% of the cases euthanasia is performed within 24 hours from the request for it, and in 50% within a week. Kimsma's conclusion is that it is medically proper for physicians to initiate the discussion on euthanasia. A patient can make an informed choice only when all options are discussed openly. It is better to open a discussion well ahead of time than for the patient to make hasty decisions, sometimes in a panic. Govert den Hartogh, a philosopher who is a member in the newly instituted Amsterdam regional committee that reviews all reported euthanasia cases in the region, expressed the same argument, while Egbert Schroten, Director of the Center for Bioethics and Health Law at Utrecht University, said that physicians can and do suggest euthanasia to their patients. After all, they know them very well and know what their needs and aspirations are.

In his very detailed response to this argument, den Hartogh maintains that in the Netherlands, the doctor is supposed to inform patients about all the relevant aspects of the decisions to be made and should never "advise" patients to actively end their life or have it ended. Nevertheless, in the case of some illnesses (e.g., cancer) in which the doctor knows from experience that the probability of severe suffering is very high, it is advisable to inform

and PAS since 1979, and that he conducts these practices when the patient is suffering and when he autonomously and voluntarily wants either of them.

18 This is a quote from Kimsma's comments on the first draft. Personal communication on September 9, 2000.

19 The Dutch Voluntary Euthanasia Society's (NVVE) internet address is: http://www.nvve.nl

the patient at an early stage that when the development of the illness takes a bad turn, euthanasia or assisted suicide can be considered. At the same time, the need for careful action must be explained and it must be clear beyond a doubt that the patient, rather than the doctor or the family, really wants to take this course of action. Even so, euthanasia will never be considered as an option for choice, but only as a measure of last resort in cases of unbearable suffering.

In the Dutch context, argues den Hartogh, providing such information at this stage does not at all convey the meaning that the doctor withdraws his support, but rather the contrary. It may provide the patient with sufficient trust to go on coping with extremely exacting conditions, and so may actually be a way of avoiding euthanasia. Furthermore, it will prevent acting in haste without carefully discussing the request and its meaning, and without allowing room for an open consultation.

Interestingly, Rob Houtepen and Heleen Dupuis, medical ethicists from Maastricht and Leiden respectively, assert that there is no basis for arguing that doctors propose euthanasia to their patients. Rob Houtepen, who never heard of such cases, thinks that physicians are restricted in this respect. Houtepen believes that doctors should not raise the subject, so as to avoid exerting pressure on patients. While the result is sometimes that the decision about what to do is delayed for too long, he feels that we should accept this. It is in the spirit of KNMG guidelines that the patient should raise the issue, not the physician. Dupuis said, in turn, that doctors are horrified by the need to perform euthanasia and that they never offer it to their patients as an option. Henri Wijsbek, a medical ethicist from Rotterdam, thinks that there might be cases in which the physician should offer euthanasia as an option, but that these cases are quite rare.

Five interviewees, Johannes van Delden who co-authored the 1990 study, G.F. Koerselman, a psychiatrist, Henk Jochemsen, a medical ethicist who objects to euthanasia on religious grounds, Chris Rutenfrans, a former law professor and currently a journalist, and Bert Keizer, a physician at a nursing home, objected to physicians' initiating a discussion of euthanasia on ethical and practical grounds. While acknowledging that many doctors conceive it as good and humane to take the initiative and raise the issue of euthanasia, they think that this might compromise the patients' voluntariness, undermine the trusted relationship between physicians and patients, and push patients to forgo life prematurely.

Keizer thinks that only the patient should initiate a request for euthanasia. He testifies that he would never put euthanasia on the menu of alternatives for treatment. It is for the patient, not the doctor, to state that all hope has gone. Keizer thinks it is awful that doctors offer death. He says he knows how to look after dying patients, and that for him there is always hope that there is something he could do. A precondition for euthanasia is the patient's request for it, evolving from his/her subjective loss of hope.

In twenty years of practice, Keizer was involved in 21–22 cases of euthanasia, on average one case per year. He does not see any ethical difference between euthanasia and PAS, but still only in the minority of cases did he perform euthanasia. As a young doctor, Keizer said he was more willing to perform euthanasia instead of PAS. Now, however, he requires the patient to show him that he/she is willing to do something about this himself/herself. Keizer would perform euthanasia only if the patient could not swallow the medication. He promises his patients that they will die within an hour and if they do not, he resorts to injecting lethal drugs to end the lingering life. It should be noted that the patient at this stage is deep in coma. This has happened in a small minority of cases. Keizer estimates that in 5 of the 21–22 cases he was required to inject a curare-like drug after the patient had taken the suicide pills. In two of the cases, euthanasia was committed because the patients could not swallow the medication.

DISCUSSION

The physician's role is commonly understood as a healing role.[20] With respect to professional ethics, talking about euthanasia upon a patient's request is different from suggesting it to the patient. By suggesting euthanasia to a patient, the physician implicitly includes euthanasia in the canon of proffered rational treatment options. In light of the professional authority that she is exercising, she thereby establishes euthanasia as a rule, and not as an exception. This conduct conflicts much more with the role of the physician as a healer than it is the case if the physician talks about euthanasia upon the patient's request. This fits into the "interpretative model" of physician-patient relationship.[21] The "interpretive model" portrays the physician as counselor, whose responsibility it is to elucidate the patient's values and to help the patient select the interventions that realize these values. The elucidation of values is complex but crucial to the principle of patient autonomy. It requires physicians to listen more than to talk.[22] Here, the physician is simply responding to an issue which the patient has raised, thereby not including euthanasia in the array of standard treatment options and implicitly emphasizing that euthanasia is an exception of the principle of physicians as healers.

The principle of physicians as healers can be perforated by the rule of allowing euthanasia in specific cases without questioning the general principle.[23] When a physician talks about the option of euthanasia upon the patient's request, we are faced with the exceptional situation in which patient's autonomy and the physician's understanding of beneficence meet and manifest in the option of euthanasia.[24] Thus, in this particular case, the healing model may be compromised in order to allow medical intervention based on a consensus between the patient and the physician which is in accordance with the principles of beneficence and patient's autonomy. But a physician should not suggest euthanasia to a patient because she would thereby establish euthanasia as a rule rather than as an exception.

Furthermore, it does not occur to Kimsma, den Hartogh and Schroten, who emphasize the autonomy and voluntariness of the patient's actions, that when euthanasia is offered, the very

20 See J.R. Peteet, "Treating Patients Who Request Assisted Suicide. A Closer Look at the Physician's Role," *Arch Fam Med*, Vol. 3, No. 8 (August 1994): 723–727; E. Rosenthal, "When A Physician Is Asked, 'Help Me Die,'" *New York Times* (March 13, 1997), pp. A1, B4.

21 See E.J. Emanuel and L.L Emanuel, "Four Models of the Physician Patient Relationship," *JAMA*, Vol. 267 (1992): 2221–2226.

22 Deborah Cook, "Patient Autonomy vs. Paternalism," *Critical Care Medicine*, Vol. 29 (2001): N24–N25.

23 Cf. Ronald Dworkin, *Taking Rights Seriously* (London: Duckworth, 1977), pp. 22–28.

24 The principles of autonomy and beneficence would also not be conflicting in a case when both, the autonomous patient and the beneficent physician, do not consider euthanasia. Conflicts arise if the autonomous patient favors an option that the physician does not see as in accordance with what she considers beneficent (of course, the content of beneficence is dependent on the particular culture). Other examples for a conflict between autonomy and beneficence may be a healthy "patient" who demands to have a good tooth extracted. The physician refuses. Or the beneficent physician offers a treatment which the patient, exercising autonomy, rejects. Those cases are not hard cases like performing euthanasia because they don't conflict with the role of the physician as a healer.

offer might undermine the patient's voluntary wishes. The patient, who trusts the long-time GP, might feel that he (most of Kimsma's euthanasia patients were men) is being condemned to death and that he is wasting the doctor's time. When all is said and done, all the physician has to offer him is death.

Interestingly, while Beusmans' patients chose physician-assisted suicide, Kimsma's patients chose euthanasia. My assumption is that patients' choices reflect their physicians' attitude. Physicians' suggestions constitute powerful influence on the patients' choices of treatment. Thus, if the assumption is correct then it may challenge autonomous decision-making by the patient, which is the precondition of ethical justification of euthanasia in the Netherlands.

The role of the physician is not to push patients to choose euthanasia. To be sure, today in the Netherlands it is impossible to argue that patients are unaware of the option. Ignorance is not a factor. Hence, physicians need to ask themselves why patients are reluctant to raise the issue. They must examine all relevant and possible answers, including the idea that the patient wishes to live despite her severe illness and medical condition. The physician should consider the consequences of what such an offer might entail for the patient's condition; for the patient's loved ones; and for the doctor-patient relationship built over the years. In a matter of life and death, caution is not only recommended, it is a must.

Den Hartogh finds it significant that in the area assigned to his regional review committee, cases of undesirable haste in the conduct of euthanasia tend to occur mainly in orthodox Protestant communities in the south, where the patient and the doctor are both reluctant to broach the subject.[25] Yet, to my mind, in the current atmosphere in the Netherlands, where it is a well-known fact that the majority of physicians support the act of euthanasia and are willing to perform it, it would suffice to make a general statement to the effect that: "I would be willing to assist you in every possible way, considering any of your wishes in order to relieve your suffering and help you cope with your condition." I find it difficult to fathom how den Hartogh does not see the compromising effect that the doctor's initiation of discussion on euthanasia might have on the relationship with the patient and on the level of trust between the two parties, as well as on the patient's mental framework.

Physicians need to remain aware of the very powerful role their recommendations can play in people's treatment choices, and of the undue ways their recommendations can influence patients. This is especially true when physicians and patients have long-standing relationships that span over decades. The challenge for physicians is to use their influence for the best purposes. Furthermore, as Peter A. Ubel warns, physicians should not make treatment recommendations that might promote their own interests against those of patients. When physicians worry that their own interests or speciality biases are influencing their recommendations, they should encourage patients to get second opinions and also try to make their biases explicit to patients.[26] It is also important to get patients to talk out loud

[25] Personal communication on August 27, 2000.

[26] Of course, second opinions may lead to contradictory recommendations. As difficult as conflicting recommendations can be for patients, these recommendations still leave patients with the option of accepting one recommendation or another, rather than feeling like they made the entire decision by themselves. When patients face difficult decisions, they are often going to learn that there is no simple right or wrong medical answer. In these situations, it is plausible to think that a series of even contradictory physician recommendations is more comforting than a series of nonrecommendations. Cf. Peter A. Ubel, "'What Should I Do, Doc?' Some

about their values before making treatment recommendations. Often, this type of conversation will make it easier for physicians to determine what recommendation is most appropriate for a patient and whether the patient is comfortable deciding what to do without receiving a recommendation.[27]

It should be noted that at the Free University Hospital, one may never decide to initiate euthanasia in the case of someone who suffers unbearably and without hope, but who has not requested euthanasia.[28] The Alkmaar Euthanasia Protocol instructs that the request for euthanasia must come from the patient himself or herself, and must be well considered, voluntary, expressed repeatedly over time, "and as permanent as possible, such as in written form or by dictation."[29] The voluntary nature of the request must be established before considering it.

CONCLUSION

The aim of this chapter was to provide account of the interviewees to the question of whether physicians should suggest euthanasia to their patients. The argument is that physicians' initiation of euthanasia might foster a sense of abandonment on the part of the patients, compromise the voluntariness of the request required by the euthanasia Guidelines, exert pressure on patients to die, and undermine seeking alternative treatments short of death, like good palliative care. Professional ethics prescribes that emphasis be put on the procedural requirement that the patient – not the physician – should raise the issue. It is disturbing to note that the majority of interviewees who are very influential in the euthanasia debate in the Netherlands discount the objections to the initiation of euthanasia.

When physicians suggest mercy killing to their patients they maneuver themselves into a situation where it is unclear whether they still act in accordance with professional ethics. The role of a physician is generally defined as the role of a healer. When the physician suggests euthanasia to a patient on her own initiative, she acts as if euthanasia was a normal treatment option, like other options that physicians offer to their patients. However, euthanasia should be handled as an exception than as a rule. It should not be included in the array of treatment options that a physician routinely offers to a patient at his/her end of life.

By not offering it to the patient but, instead, talking about it only upon the patient's request, the physician implicitly acknowledges the exceptional character of the case, and the conflict of performing euthanasia and her healing duty. In particular cases, she is willing to do it, because the patient requests it and the individual physician is able to see the request in accordance with her understanding of beneficence.[30]

Psychologic Benefits of Physician Recommendations," *Archives of Internal Medicine*, Vol. 162, No. 9 (May 13, 2002).

27 *Ibid.*

28 Gerrit K. Kimsma and Evert van Leeuwen, "Comparing Two Euthanasia Protocols: The Free University of Academic Hospital Amsterdam and the Medical Center of Alkmaar," in David C. Thomasma, Thomasine Kimbrough-Kushner, Gerrit K. Kimsma, and Chris Ciesielski-Carlucci (eds.), *Asking to Die* (Dordrecht: Kluwer Academic Publishers, 1998), p. 119.

29 *Ibid*, p. 124.

30 Cf. T.L. Beauchamp and J.F. Childress, *Principles of Biomedical Ethics* (New York: Oxford University Press, 1989), 3rd edition.

That is to say, in sum, that euthanasia is an option in the Netherlands but there is a difference between a patient who initiates the discussion about it upon realizing that the end of life is nearing, and a physician who is putting it in the menu of treatment options. The physician should be convinced that euthanasia is, indeed, the wish of the patient and that he/she is sure to have it, sure enough to find the energies within him/herself to raise it. In this context I should emphasize the importance of open and candid communication between patients and their physicians. In *Asking to Die,* a physician testified that he found that when patients know euthanasia can be an option for them, they often keep silent about it. Open discussions help them face their disease because they know that if pain becomes unbearable and euthanasia is really necessary, "I am going to help them." Knowing euthanasia is an option gives the patients comfort and they do not necessarily have to act on it. Satisfied to have this support, most of the time they do not even talk about euthanasia anymore and they go on to die of natural causes.[31]

Parenthetically, it is worth noting that in an American study conducted by Ezekiel Emanuel and his colleagues,[32] oncology patients were interviewed to provide empirical data of patients' attitudes and practices related to euthanasia and physician-assisted suicide. Oncologists and members of the general public were also interviewed to compare their responses with those of the oncology patients. It was found that 53% of oncologists, as compared with 37.2% of patients and 44.4% of the general public, thought that discussions between patients and physicians on "end-of-life care that included explicit mention of euthanasia or physician-assisted suicide" would reduce patients' trust in the physician. By contrast, 41.6% of patients, 32.8% of the general public, and only 15.6% of oncologists thought that such discussions would increase patients' trust in the physician. Patients with depression and psychological distress were significantly more likely to feel that discussions which included explicit mention of euthanasia or physician-assisted suicide would increase trust in their physician, whereas patients with pain believed that such discussions would not increase trust.[33]

31 "Annie Asked, 'Are You Going to Help Me?,'" in David C. Thomasma, Thomasine Kimbrough-Kushner, Gerrit K. Kimsma, and Chris Ciesielski-Carlucci (eds.), *Asking to Die, op. cit.*, p. 279.

32 Leenen wrote in his comments that the conclusions of this study are not applicable to the Netherlands because the United States lacks the family-doctor system "as we have it." Letter dated July 25, 2000.

33 Ezekiel J. Emanuel, Diane L. Fairclough, Elisabeth R. Daniels and Brian R. Clarridge, "Euthanasia and Physician-assisted Suicide: Attitudes and Experiences of Oncology Patients, Oncologists, and the Public," *Lancet,* Vol. 347 (June 29, 1996), p. 1808. The Assisted Suicide Consensus Panel asserts that physicians should not encourage patients to hasten death, even when practicing in jurisdictions that allow assisted dying. Cf. James A. Tulsky, Ralph Ciampa and Elliot J. Rosen, "Responding to Legal Requests for Physician-Assisted Suicide," *Annals of Internal Medicine,* Vol. 132 (March 21, 2000): 494–499. For further deliberation, see Timothy E. Quill, "Initiating End-of-Life Discussions with Seriously Ill Patients: Addressing the 'Elephant in the Room,'" *JAMA,* Vol. 284, No. 19 (November 15, 2000): 2502; Susan D. Block, "Psychological Considerations, Growth, and Transcendence at the End of Life," *JAMA,* Vol. 285, No. 22 (June 13, 2001): 2898; Maria E. Suarez-Almazor, Catherine Newman, John Hanson and Eduardo Bruera, "Attitudes of Terminally Ill Cancer Patients About Euthanasia and Assisted Suicide: Predominance of Psychosocial Determinants and Beliefs Over Symptom Distress and Subsequent Survival," *Journal of Clinical Oncology*, Vol. 20, Issue 8 (April 2002): 2134–2141.

Of course, it is not easy to compare between the healthcare conditions of the USA with those of the Netherlands. At the same time, note the relatively high percentage of oncologists who believe that it is unwise to explicitly mention euthanasia and PAS in discussions with patients. The patients have no clear idea about the issue and present divided opinions. Interestingly, depressed patients and patients with psychological distress saw value in explicit mentioning of euthanasia or physician-assisted suicide. For them, euthanasia and PAS should be considered as viable solutions. Instead, some therapy is advisable to treat the depression and help them find some meaning in life.

CHAPTER 9

BREACHES OF THE GUIDELINES

CONSULTATION

The notification procedure states the following criteria for good consultation: the consultant should not work in the same practice or be a co-attending physician, should see the patient, and should consult a psychiatrist if the patient has a psychiatric illness. The consultant should draw conclusions concerning the patient's condition and estimated life-expectancy, the possibility of alternative treatments, and whether the patient's request is voluntary, well-considered and persistent.[1]

My next question explored the issue of consultation, as follows: The physician practicing euthanasia is required to consult a colleague in regard to the hopeless condition of the patient. Who decides who the second doctor will be? I also asked about what happens in small rural villages where it might be difficult to find an independent colleague to consult. One prosecutor told John Keown that in the countryside there are towns with only two or three doctors. He therefore asked rhetorically: "What's the use of asking one of those two or three to judge the handling of a euthanasia case by the other one? How objective can that be? I don't see it."[2]

The Dutch movie *Death on Request,* broadcast on Dutch television in October 1994, showed that the doctor performing the euthanasia called a colleague to consult with him about his ALS patient.[3] It is unclear why Dr. Wilfred van Oijen picked this specific consultant. Was it because of his particular field of expertise or because the physician knew this doctor and assumed that he would back his decision without too many questions? From the tone of the discussion, it seemed that the two doctors knew one another quite well and were on friendly terms. What worries me is that the requirement to consult could become a "dead dogma",[4] used only to fill the reports, and that, in essence, one hand simply washes the other: you approve euthanasia for my patients, and I will approve it for yours. Obviously, a doctor who approves of euthanasia would not call a colleague that is against it or is hesitant about the practice.[5] Indeed, one study showed that the consultant was nearly always a partner

1 Breje D. Onwuteaka-Philipsen, Gerrit van der Wal, Piet J. Kostense and Paul J. van der Maas, "Consultation with another Physician on Euthanasia and Assisted Suicide in the Netherlands," *Social Science and Medicine,* Vol. 51 (2000), p. 430.

2 John Keown, "The Law and Practice of Euthanasia in the Netherlands," *The Law Quarterly Review,* Vol. 108 (January 1992): p. 68.

3 *Death on Request,* IKON, Interkerkelijke Omroep Nederland, Postbus 10009, 1201 DA Hilversum. I thank IKON for sending me a copy of this film. For deliberation and critique of the content of this film, see Herbert Hendin, *Seduced by Death* (New York: W.W. Norton, 1997), pp. 114–120.

4 Cf. John Stuart Mill, *Utilitarianism, Liberty and Representative Government* (London: J.M. Dent, 1948), chapter 2 of *On Liberty*, p. 95.

5 Jochemsen thinks that the patient was pushed to ask for euthanasia. His wife made most of the talking and supported the euthanasia decision. The patient had a limited role in the conversations. And van Oijen was

in the practice or a locum. At least 60% of the "independent consultants" giving the second opinion already knew the patient before the consultation. In only 5% of the cases did the family doctor seek a second opinion from a doctor that he did not know personally.[6] Another study showed, unsurprisingly, that almost all consultants regarded the request of the patient to be well-considered and persistent, conceded that there were no further alternative treatment options, and agreed with the intention to perform euthanasia or assisted suicide. In general, the GPs did not need to change their views or plans following the consultation.[7]

Consultation takes place in about 99% of reported cases of euthanasia and assisted suicide (to reiterate, only 41% of cases are reported). It is estimated that consultation takes place in 37% of unreported cases. In 88% of cases, the consultant has seen the patient.[8] Physicians mainly consult colleagues from their own specialty. Recent research shows that familiarity and accessibility are very important factors in the choice of the consultant. Half of the physicians who had served as consultants more than once had previously been consulted in the most recent case of the same physician. In 24% of these cases, the treating physician and the consultant had previously acted as consultants for each other. Physicians who previously consulted or had been consulted by the same physician agreed more often with the intended euthanasia or assisted-suicide than physicians who did not (90% vs. 80%, respectively).[9]

The interviews revealed sharply contrasting and contradictory opinions on the matter of consultation. I suspect that not all of the interviewees were completely candid in their answers, again possibly because they were "protecting the system" and viewed me with suspicion as a "foreigner."

John Griffiths says that there are problems in the consultations of doctors with their colleagues. This may be particularly true in rural areas, where it can be difficult to get hold of a colleague, especially an "independent" doctor, as doctors in rural areas are often members of the same substitution group. In Griffiths' opinion, the consultation requirement should be adhered to more strictly than now appears to be the case. However, the complexities of concrete situations require a rule that can be applied in a flexible and casuistic way, something that is difficult in the context of criminal enforcement. Currently the courts are rather lenient with doctors who do not comply, but the regional assessment committees seem to be trying to give the requirement more teeth.[10] In this context, Jacqueline

presented as a hero. Jochemsen believes it is obscene to perform euthanasia in front of the TV-camera and in addition gain reputation for this.

6 G. van der Wal, J.Th.M. van Eijk, H.J.J. Leenen and C. Spreeuwenberg, "Euthanasia and Assisted Suicide. II. Do Dutch Family Doctors Act Prudently?," *Family Practice,* Vol. 9, No. 2 (1992), pp. 113, 115.

7 Bregje Dorien Onwuteaka-Philipsen, *Consultation of Another Physician in Cases of Euthanasia and Physician-assisted Suicide* (Amsterdam: Vrije Universiteit, 1999), Thesis, pp. 29, 31. Interestingly, Onwuteaka-Philipsen *et al* report in another study that in 17% of cases of intended euthanasia and PAS the consultant advised against the practice and that in almost all of these cases the attending physician acted in accordance with the consultant's judgment. Cf. Breje D. Onwuteaka-Philipsen, Gerrit van der Wal, Piet J. Kostense and Paul J. van der Maas, "Consultation with another Physician on Euthanasia and Assisted Suicide in the Netherlands," *Social Science and Medicine,* Vol. 51 (2000), p. 436.

8 Bregje D. Onwuteaka-Philipsen, Gerrit van der Wal, Piet J. Kostense and Paul J. van der Maas, "Consultants in Cases of Intended Euthanasia or Assisted Suicide in the Netherlands," *MJA,* Vol. 170 (1999): 360–363.

9 *Ibid.,* pp. 360–363.

10 Henri Wijsbek reiterated this point of lenient courts saying he did not know of any prosecutions for lack of consultation, and that consultation should be "observed and complied closely." Leenen, on the other hand,

M. Cuperus-Bosma *et al* examined the Minutes of the Assembly of Prosecutors General and note that if all requirements for accepted practice were met, except consultation, the physician was not prosecuted but the case was usually referred to the Health Inspector. However, if there were doubts about other requirements for accepted practice being met, an inquest was held.[11]

Furthermore, according to Griffiths, some of the prosecuted cases involved a doctor who consulted another doctor, and then the consulted doctor performed the euthanasia, rather than the *first* doctor who asked for the consultation. However, the Guidelines say that the physician who first recommended euthanasia, not the consulted doctor, should perform the euthanasia. When this "role reversal" occurs, the case may not be reported for fear of prosecution. Griffiths added that it is wrong to suppose that all unreported cases are unjustified.

Griffiths estimated that 10% of physicians in the Netherlands principally oppose the practice of euthanasia, and that a further 6% would not perform euthanasia themselves, but refer patients who ask for it to another doctor.[12] Griffiths further stated that there are indications from the regional assessment committees that consultation by specialists is sometimes quite inadequate, being performed over the phone or by a busy specialist who stops by a hospital ward and notes on the patient's status sheet that he agrees with the attending physician. He argues that consultants should always see the patient, but that the prosecution and the courts do not regard this as an absolute requirement. The Supreme Court should broaden the requirement of consultation in person to all patients, and not limit it to psychiatric patients only.[13] The nursing staff should also be included in the euthanasia discussions. In cases of euthanasia performed at the patient's home, the patient usually has home nursing care, and the on-site nurse should be included in the decision-making process.

Sjef Gevers explained that until 1995, consultants did not need to see the patients. The Dutch Medical Association Euthanasia Guidelines of 1995 changed the picture, requiring that the consultant be an independent colleague, rather than part of the doctor's group, and

wrote that it is incorrect to say that the courts are very lenient towards lack of consultation. Letter dated July 25, 2000.

11 Jacqueline M. Cuperus-Bosma, Gerrit van der Wal and Paul J. van der Maas, "Physician-assisted Death: Policy-making by the Assembly of Prosecutors General in the Netherlands," *European J. of Health Law*, Vol. 4 (1997), p. 232.

12 In his book, Griffiths writes that twelve percent of Dutch doctors are principally unwilling to perform euthanasia and that most of them would refer a patient requesting it to another doctor. See John Griffiths, Alex Bood and Heleen Weyers, *Euthanasia and Law in the Netherlands* (Amsterdam: Amsterdam University Press, 1998), p. 253. According to van der Maas *et al*, 9% of all physicians would never perform euthanasia and assisted suicide but would refer patient seeking it to another physician. 3% would never perform the practices or refer patients. Cf. Paul J. van der Maas, Gerrit van der Wal, Ilinka Haverkate *et al*, "Euthanasia, Physician-Assisted Suicide, and Other Medical Practices Involving the End of Life in the Netherlands, 1990–1995," *New Eng. J. of Med.*, Vol. 335, No. 22 (November 28, 1996), p. 1702. Compare to physicians' attitudes in other countries: B.J. Ward and P.A. Tate, "Attitudes among NHS Doctors to Requests for Euthanasia," *British Medical Journal*, Vol. 308 (May 21, 1994): 1332–1334; Diane E. Meier, Carol-Ann Emmons, Sylvan Wallenstein *et al.*, "A National Survey of Physician-Assisted Suicide and Euthanasia in the United States," *New Eng. J. of Med.*, Vol. 338, No. 17 (April 23, 1998): 1193–1201; Ethics and Human Rights Committee, American College of Physicians-American Society of Internal Medicine, "Physician-Assisted Suicide [Position Paper]," *Annals of Internal Medicine*, Vol. 135, No. 3 (August 7, 2001): 209–216.

13 Griffiths added that the physician is supposed to discuss the matter with the patient's family and, in his opinion, should be required to explain in writing if this is not done. If the patient does not wish the family to be included in the deliberations, then the doctor should be required to have the patient put that refusal in writing.

must talk with the patient himself and be informed of the patient's medical situation. Consultation over the phone or by looking at the patient's file is insufficient. However, the courts do not insist that the consultant see the patient. Following the *Chabot* precedent, consultation is required in person only in psychiatric cases.

Several interviewees[14] explained that in hospitals the general practice is to consult the whole medical team, including nurses, not just another physician. Thus in hospitals, consultants always see the patients; examining their medical files is considered insufficient. In nursing homes, the standard procedure is to invite a consultant from another nursing home. As for GPs, many physicians have a trusted colleague whom they always consult in euthanasia cases. It was noted that it is important that the consultant not be from the GP's medical team or someone who fills in for the doctor on weekends. However, often GPs consult colleagues on their own team. The consultant is perceived to be independent because he or she is not directly involved with the patient, but of course that is not total independence from the perspective of the best interest of the patient. The common view is that the physician needs to hear and see the patient, examine him, feel him, listen to what the patient wants.

On the other hand, Ron Berghmans and A. van Dantzig do *not* think that finding an independent doctor is a major problem. The Netherlands is a small country, and it is possible to find a consulting doctor who does not belong to the same medical team who would be willing to travel to see the patient. Berghmans thinks the GP and the consultant might have other shared interests, but that they would not compromise the independence requirement. With regard to consultation over the phone, Berghmans contends that in the past too much respect was granted to maintaining privacy in physician-patient relationships, even to the extent of allowing consultation over the phone. This picture is now changing, and the new law requires that the consultant see the patient.

Paul van der Maas explains that "real consultation" means consultation with a colleague who is an expert in the field and who is able to verify that there are no available alternatives for treatment. The consultant should also verify that the patient really wants euthanasia, and that the decision-making process did not involve problems of transference and counter-transference between doctor and patient.[15] Van der Maas maintains that he and his team train consultants to see the patient and examine his condition first hand. Similarly, van Dantzig argues that consultation involves seeing the patient, determining the motive for the wish to die, and exploring avenues of treatment. Seeing the patient is required to verify that euthanasia is the only solution, and the most desirable solution.

Frank Koerselman was consulted in the past and objected to the practice. He testified that he felt pressure from his colleagues to sign the documents approving the euthanasia decision. At some point, his colleagues gave up on him and stopped consulting him. Koerselman offered to serve as an expert witness in one court trial, but the court was not interested in hearing his expert testimonial.

[14] Bert Thijs, Rob Houtepen, Arie van der Arend, Jaap Visser, Ruud ter Meulen, and Henk Jochemsen.

[15] Freud recognized that doctors' unconscious has an impact on their relations with patients. Counter-transference has been defined in the psychoanalytic literature as reactions in the therapist engendered by the patient. Cf. Jay Katz, *The Silent World of Doctor and Patient* (New York: The Free Press, 1984), p. 147.

I questioned how much time is needed for consultation. Van der Arend thought that a totally independent physician is unable to evaluate the condition of the patient within the customary 1/2 hour or hour of consultation. Ideally, the consulting physician should meet the patient several times. If there is only one short meeting, there may be a lack of communication. Van der Arend advised having three different meetings before the consultant writes the report.[16]

George Beusmans and Gerrit Kimsma do not share this view. According to Beusmans, after several meetings with the patient, he asks for the patient's request for euthanasia in writing. At a certain moment, when the patient says: "I can't deal with the suffering, you can do it," Beusmans asks: "When?" This discussion takes place when the patient is in the final phase of life, having only about two weeks left. When the patient insists that he wants euthanasia, Beusmans arranges for a colleague to come and see the patient. At this stage, Beusmans also contacts a pharmacy to arrange for the lethal drugs. The colleague is a General Practitioner with whom Beusmans does not work. Beusmans has two colleagues with whom he cooperates on euthanasia matters, and these two colleagues also ask Beusmans to serve as a consultant for their euthanasia cases. The consultant will see the patient, speak with him, and decide whether it is necessary to perform euthanasia. The consultant usually signs the papers after thirty minutes of conversation with the patient. Beusmans thinks thirty minutes is enough to verify that the patient qualifies for euthanasia.

Gerrit Kimsma also insists that the consultant see the patient. He thinks 1/2 hour is enough when the consultant comes prepared with all the pertinent information.[17] The consultant reads the patient's medical records, sees the patient, asks for the patient's view on his condition, and checks whether the patient knows why the consultant came. The consultant needs to see that the patient is of sound mind and is requesting euthanasia without pressure. He is required to verify that the Guidelines have been fulfilled and that Kimsma's euthanasia decision was correct.

Kimsma testifies that he consults a colleague for whom he covers during the weekends. In his opinion, the independence requirement is not compromised because the main concerns are to examine the issues of transference and counter-transference, and to determine that the GP has arrived at the euthanasia decision without pressure and without identifying with the patient to the point of obscuring his own medical judgment.

My interviews also included a meeting with the de B. family who experienced the euthanasia decision-making process. K was a cancer patient who knew that death was inevitable. He could not adequately digest food and was very weak; he suffered great pain and consumed large doses of pain medication. K felt that his life had no quality and filled out the papers he had obtained from the Dutch Voluntary Euthanasia Society, in which he expressed a wish to die. He reiterated his request to his personal doctor and at a later stage entered into a state of unawareness for a few days. The meeting with the consultant to approve the GP's decision was scheduled ahead of time, and on that day the consultant arrived an hour after K woke up. K was in a good mood and did not believe that he had slept for four days. The GP told K that he had arrived to discuss K's euthanasia decision, and K

16 As a nurse, van der Arend is dissatisfied that nurses are not represented on the committees. In his mind, it would be better to have a balance of ideas before making the euthanasia decision by including nurses and independent physicians, and by following the rules of carefulness in detail.

17 Van Leeuwen also thinks half an hour is sufficient for the purpose of consultation.

stated he did not believe the situation was that bad; he thought his family and the physicians had made this up. The consultant talked with K about euthanasia, but K found it difficult to comprehend why the consultant wanted to discuss euthanasia with him since he had had such a good sleep and was feeling quite happy. Clearly, the family testified, K was not ready for euthanasia though K still backed his euthanasia decision. I asked what the consultant decided after this confusing episode and was told that the consultant arrived again later, when K's condition had deteriorated yet again, and confirmed the decision for euthanasia.

Though I felt that K's family acted sincerely in a *bona fide* manner, this episode is disturbing and demonstrates the intricacy of this issue. It is unclear why the GP and the consultant arrived that day. If K had been unconscious for four days, the GP should have been aware of this. Surely, the consultant could not fulfill his responsibility if the patient was unaware and unable to communicate. The consultant here was not satisfied by the first visit, during which the patient clearly wanted to live, and felt an obligation to visit the patient again. Additionally, it must have been quite a blow for K to see his beloved family and the physicians, including his trusted GP, around his bed discussing his mercy killing at a time when he felt well.

Bert Keizer tells the disturbing story of a cancer patient who arranged to end her life, but during the last days became increasingly muddled. On the evening of her death, when she heard the doorbell, she let the doctor in, greeting him with some bewilderment: "And what brings you here tonight, doctor?" The doctor and the other people present at her home refreshed her memory and later that evening the patient did take her dose. Before the doctor left he asked the patient's daughter: "This *is* what Mother wanted, isn't it?"[18]

Many interviewees spoke about the new Support and Consultation of Euthanasia in Amsterdam (SCEA) project that began in Amsterdam and became a national Dutch project (SCEN).[19] In 1997, the SCEA project was initiated to provide all GPs working in Amsterdam with a support group of about twenty especially trained GPs for consultation or advice on euthanasia and PAS. The purpose was not only to make it easier for GPs to find an independent and knowledgeable consultant, but also to make the consultation more professional.[20] SCEN provides a telephone service that enables doctors to speak with trained, professional staff who can offer practical medical or legal advice or arrange a formal consultation. Staff will follow a consultation protocol, and a written record of consultations will be kept.[21] Physicians are required to contact SCEN consultants before they perform euthanasia, in order to make consultation as effective as possible. Van der Wal says that most doctors do not like the idea that they do not select the doctor themselves, especially at the time when euthanasia was officially illegal. Physicians preferred to consult with someone they knew. Gerrit Kimsma saw no problem in choosing the consultant himself or, for that matter, in allowing every GP to choose his/her consultant. He said that there was good faith among physicians.

18 Bert Keizer, *Dancing with Mister D* (London: Black Swan, 1997), p. 117.

19 Interviews with Evert van Leeuwen, John Griffiths, J.K. Gevers, Gerrit van der Wal, Jaap Visser, H.J.J. Leenen, Henk Jochemsen, Gerrit Kimsma, Paul van der Maas, Govert den Hartogh and Johannes van Delden.

20 Bregje Onwuteaka-Philipsen, *Consultation of Another Physician in Cases of Euthanasia and Physician-assisted Suicide* (Amsterdam: Department of Social Medicine, Vrije Universiteit, 1999), Doctoral Thesis, p. 91.

21 Tony Sheldon, "Netherlands Sets Up Euthanasia Advisory Body," *British Medical Journal*, Vol. 318 (February 6, 1999): 348.

Four of the interviewees (van Leeuwen, Kimsma, van Delden and den Hartogh) are members of the newly established regional committees whose role is to review euthanasia cases and see that the rules of carefulness are observed. In November 1997, the Secretaries of Justice and of Healthcare, Well Being and Sports published their intention to inaugurate five regional review committees to supervise physicians in actively ending the lives of their patients. The five regions are (1) Groningen, Friesland and Drenthe; (2) Overijssel, Gelderland, Utrecht and Flevoland; (3) North Holland; (4) South Holland and Zeeland; (5) North Brabant and Limburg. These committees have been functioning since November 1998 and retrospectively evaluate the reported cases of euthanasia and physician-assisted suicide. The committees' members consist of a lawyer (who serves as chairperson), a physician, and an ethicist, and their responsibility encompasses all cases in which a voluntary request has been made by a competent patient. Cases of physician-assisted death without such a request are sent directly to the Office of the Prosecution. The primary goal of having regional committees is to evaluate the prudence of the practice of physician-assisted death, with the intent of public control of a highly sensitive medical practice and moral issue. The secondary goal is to increase the number of reported cases and thus make public control more effective.[22] In short, the role of the committee is to examine whether the physicians observe the rules of caution, including reporting.

Evert van Leeuwen and Govert den Hartogh provided detailed insight about the respective regional committees on which they are members. Each month, van Leeuwen's committee examines 50 cases of euthanasia and physician-assisted suicide that took place in North Holland. In turn, den Hartogh's committee reviews 36 cases per month. Other regions have 35 cases on average. Each regional committee meets once a month, and each of the committee members reads all cases before the meeting. They try to reach a consensus on each and every case. Van Leeuwen, den Hartogh and their colleagues on the committee review the files in a given case and verify that the physician made a careful judgment according to the Guidelines. They assess the durable wish of the patient, the patient's suffering, the GP's consultation with a colleague, and the use of proper drugs to perform the euthanasia.

Thus, the committee provides moral support to physicians who conduct euthanasia in a moral way. If euthanasia has not been done according to the Guidelines, then the committee asks the physician to provide clarification and further information. Each report should contain a declaration by the patient, the physician's report, a statement from the consultant, and the coroner's statement. Sometimes there is also a letter from the family.

Van Leeuwen emphasized that the committee is not a prosecutorial body. Its aim is to convince physicians to report. Van Leeuwen explained that the KNMG envisioned a body of control between the law and the practice. Until the early 1990s, the police checked every incident of unnatural death. In some regions, the police arrived at both the home of the physician and the home of the patient; in other regions, the police arrived at the physician's home only. This was very disturbing, so those visits were stopped in the early 1990s.

In van Leeuwen's opinion, euthanasia is not only a medical act; it is an extraordinary act that physicians should report to the public. However, there are physicians who consider it a

[22] Evert van Leeuwen and Gerrit Kimsma, "Problems Involved in the Moral Justification of Medical Assistance in Dying: Coming to Terms with Euthanasia and Physician Assisted Suicide," in R. Cohen-Almagor (ed.), *Medical Ethics at the Dawn of the 21st Century* (New York: New York Academy of Sciences, 2000): 157–173.

private matter between themselves and their patients. The regional committees are attempting to change this view. They are also attempting to persuade physicians that if euthanasia is performed properly, then there is no reason to fear prosecution. Members of the committees write letters to physicians explaining the need for reporting and how they should perform euthanasia. However, the letters are sent to physicians who report, not to those who don't, as the committee gets their names from the reported files.

The committee's verdict on each euthanasia case goes to the District Attorney's (DA) Office, where the prosecutor checks whether the committee examined the case thoroughly. There have been a few cases in which the DA disagreed with the conclusion of van Leeuwen's committee. Den Hartogh testified that there was no single case in which the prosecution overruled the decision of his committee. Under the new law, the regional committees review the reported euthanasia cases. Lawyers object to granting the committees the power to decide whether or not to prosecute because there is only one lawyer on each committee. Van Leeuwen expects that there will be discussions on this issue in parliament and that ultimately this power will not be granted. He believes that the parliament will seek a way for the DAs to retain their freedom to prosecute.

Most of the reported cases were cancer patients (90% of the cases reported to den Hartogh's committee; 80–90% of the cases reported to van Leeuwen's committee). Den Hartogh said that cancer patients are the accepted group for euthanasia and speculated that doctors might not report euthanasia performed for non-cancer patients because they fear the committee might consider this conduct as unusual and might launch an inquiry into the matter. This is an interesting speculation. Is it the case that most euthanasia cases involve cancer patients or, alternatively, that physicians who provided mercy killing to non-cancer patients did not report, and hence contributed indirectly to the data that associate euthanasia with cancer patients? This is a difficult and interesting question, which requires further empirical research and analysis.

Van Leeuwen's committee had reviewed some 300 cases by the time of his interview, and in most of these cases the Guidelines had been observed. In a small number of cases (between four and eight), the requests were very clear, the physicians could do nothing to help, and the patients were suffering, but were not on the verge of death; they still had four to six months to live. Van Leeuwen felt that these were cases in which palliative care could have helped. This issue is something that still needs to be explored and developed. Govert den Hartogh testified that the problematic cases usually involved consultants who were not truly independent. Sometimes the patient did not form what den Hartogh terms a "categorical request" for euthanasia, or the practice was conducted too early. Nevertheless, the committee only "on occasion" asked physicians to clarify their actions, and only "rarely" concluded that the physician's actions were not carefully taken.

In his detailed comments on the first draft of this study, den Hartogh elaborated his explanation by saying that the cases in which his committee asked for more information and/or clarification, were not cases involving a breach of the Guidelines. Rather, these were cases in which the information provided (by the doctor, the consultant or both) was insufficient to make a reliable judgment on the issue. Such probing questions were asked in 15–20% of the cases. Den Hartogh added that other committees have done so less frequently.

In 1–2% of the cases, members of the committee were not satisfied with the replies and invited the doctor for an interview. In one case, the committee decided after the interview, "the patient had not made a relevant request, so the committee was not competent to decide the case, and the report was sent to the public prosecutor." In three or four cases, the committee had some doubts concerning the presence of unbearable suffering, "but having interviewed the doctor we finally decided that his action had met the criteria." In three cases, the committee found that the requirement of independent consultation had not been satisfied. In three cases, the committee's final judgment was that the doctors had acted carefully on the whole, but that during the procedure some mistakes had been made, either by themselves or their colleagues, requiring the attention of the health care inspection agency.[23] These last cases have all been scrutinized by the public prosecutor, which did not lead to actual prosecution. The committee never recommended prosecution; it only recommended investigation by the health care inspection agency.[24]

Sometimes the committee saw from the report that the request for euthanasia, the consultation with another doctor, and the act of euthanasia had all been performed on the same day. Den Hartogh explained that this happened when the patient was suffocating and suffering severely. Ordinarily, such rapid decision-making should not take place. Nevertheless, these cases constituted, in den Hartogh's view, "unavoidable exceptions."

Evert van Leeuwen testified that his committee did not review even one incident of consultation conducted over the phone. He considers it is essential for the consultant to see the patient, to verify that he or she is competent and acting upon free will, and to review the patient's medical condition, by both physical examination and also examination of the medical files.

Govert den Hartogh explains that doctors who do not consult a colleague do not report to the regional committee. KNMG advises consultants to see the patients, and the consultants have adhered to this requirement. The reports he reviewed said the consultation lasted 1 to 2 hours, but den Hartogh testified that he was unsure about this. He is certain that the time for consultation is often shorter, especially in hospitals. Den Hartogh maintained that doctors in some islands in the south might find it difficult to find a consultant. In his comments on the first draft he wrote that in August 2000 his committee reviewed a case of a doctor from one of those southern islands with an orthodox Protestant majority, in which the doctor had consulted his own associated partner. The physician explained that he had tried to find another consultant but without success. Thus, such cases do occur, though probably rarely. In the reported cases, there is always a consultant who has seen the patient; if there are still cases in which all consultation has been done over the phone, they are never reported.[25]

Den Hartogh further wrote in his comments that one unfortunate side-effect of the fact that the rules for justifiable euthanasia are court-made and rely on the defense of necessity is that the matter of consultation has not been given sufficient attention. As a result of KNMG-policy and of the growing involvement of the government in the assessment of acts of euthanasia and assisted suicide, this matter has gradually been improving. In hospitals the report is often nothing but a short written note on the patient's state. But there is evidence that the SCEN project already has had good effects on the quality of both consulting and reporting. Den Hartogh believes that the training of SCEN consultants and of doctors generally will be far more effective in shaping the Dutch practice than any possible form of legal regulation.

23 This is an agency, independent of the public prosecution. Its activities may lead to disciplinary law trials.
24 Den Hartogh's personal communication on August 27, 2000.
25 Personal communication on August 27, 2000.

Since the installation of the regional review committees, the requirement is that the consultant should be independent, that he/she should see the patient in person, and consider both the character of the request and the nature of the suffering. Den Hartogh clarifies that that doesn't mean that a doctor who failed to consult a colleague at all, or failed to consult an independent one, could not appeal to necessity. He/she can make that appeal even when the new law has formalized the new requirement.[26]

Johannes van Delden is a member of a third regional committee. His response to my inquiry about his work was far more reserved than the responses of van Leeuwen and den Hartogh. Van Delden said that almost all reported euthanasia cases "had something in writing," but he is "not allowed to say how many cases there were."[27] The documents, *inter alia,* clarified that the patient had made the euthanasia request. Van Delden maintained that it is too early to judge the regional committees' effectiveness. He explained that if the committees were too harsh on the doctors, then they would fail to comply with reporting requirements. On the other hand, if the committees were too lenient, then their work would have no real purpose. Thus, the committees must attempt to preserve a delicate balance in their work. They tend to keep the process outside the realm of criminal law and to emphasize educating the doctors. If it appears that a doctor did not follow the Guidelines, then discussions will be held with him or her and, if required, with the consultant as well, explaining what was lacking and how their practice of euthanasia should be improved.

Van Delden maintains that the consultant should see the patient for one hour after examining the patient's medical files and speaking with the GP, inquiring whether the doctor explored other medical alternatives prior to the euthanasia decision. The consultant is required to explain his reasoning; simply writing "I agree" on the form is insufficient. Van Delden's committee asks for detailed explanations. The role of the committee is also educational, explaining that the reports should be informative. According to van Delden's testimony, there was only one incident in hundreds of cases reviewed by his committee in which the GP consulted a colleague over the phone. The committee reported the case to the medical inspector. Van Delden does not think there is any problem with consultation in rural areas. Most doctors are willing to be involved in the practice of euthanasia, and it is not difficult to find an independent doctor.

Many interviewees[28] oppose the practice of looking at the medical files in lieu of examining the patient, thinking that the consideration of unbearable suffering should be closely examined, and physicians cannot verify that by only looking at the files.

One interviewee spoke of Sippe Schat, a physician who was viewed as "a God in his village," who did not consult colleagues, and who did everything alone without consultation. Eventually he was prosecuted and found guilty for not consulting a colleague prior to performing euthanasia (District Court, Leeuwarden, April 8, 1997).[29]

26 Personal communication on August 27, 2000.

27 In his comments on the first draft of this chapter, van Delden explained that his hesitation to disclose numbers at that point in time derived from the fact that these numbers were not public yet. Personal communication on August 4, 2000.

28 Van Leeuwen, Griffiths, Gevers, Thijs, van der Arend, Houtepen, Jochemsen, den Hartogh, van Delden, Visser and van der Maas.

29 Griffiths *et al.* argue that the facts found by the District Court, involving multiple and serious failures to conform to the requirements of careful practice, seem to call for a serious medical disciplinary measure, perhaps revocation of the license to practice medicine. See John Griffiths, Alex Bood and Heleen Weyers, *Euthanasia and Law in the Netherlands, op. cit.*, p. 293, note 56.

Egbert Schroten said that, to the best of his knowledge, most doctors consult their colleagues, and at least until the early 1990s the consultation was done over the phone. When I asked whether this is sufficient to warrant euthanasia, Schroten answered that doctors apparently think they can approve euthanasia without seeing the patients, believing they have enough information to decide the matter.

These testimonies are alarming. I question whether it is possible to conduct a reliable consultation over the phone. It should be obligatory to see the patient, to examine him or her, to confirm that the patient freely wishes euthanasia, and that all options for treatment were exhausted before resorting to medical killing. At first I was astonished by the interviewees' relaxed tone while speaking about consultation by telephone. In turn, they were somewhat surprised to see my alarm.

H.J.J. Leenen does not share my alarm. He explains that during the 1980s, consultation was often conducted over the phone. Euthanasia was regarded as any other medical practice. Leading decision makers and policy consultants, among them Leenen himself, said that euthanasia was, is and should remain an exception. Consequently, a view emerged that euthanasia is not like other medical procedures that could be consulted about over the phone. Physicians now make their decisions by looking at the medical files or by meeting patients in person. Although euthanasia is an exceptional medical procedure, Leenen does not think the consultant should always see the patient. In his view, consultation in person is often unnecessary. Leenen doesn't agree with the Dutch Medical Association (KNMG) 1995 directives that consultants see the patients, because most of those asking for euthanasia are dying from cancer, and their evaluations can be done by reviewing their medical files. Many patients' families regard the consultation requirement as a beaurocratic stupidity, a redundant control mechansism. Leenen agrees that consultants need to see psychiatric patients, but feels there is no such need in what he terms to be "clear somatic cases."[30] He states that he trusts doctors and the experience he had working with doctors for the past twenty-five years has been positive: "Doctors are morally decent and competent people." Leenen criticizes them for not spending enough time with their patients, "but their intentions are good." Leenen maintained that consultation was a problem in small villages, but in May 1999, following SCEA, the government initiated the organization of consultation teams all over the country. Consultants will travel to small villages to examine medical files and to see patients. Hospital specialists are required to examine the files. The scheme is not fully operational yet, and time will tell to what extent it will succeed; Leenen thinks the consultation mechanism has gradually improved.

Heleen Dupuis contested the views of most of her colleagues. Unlike Leenen, she thinks the wish of the patient and his/her medical condition need to be confirmed by a second opinion; hence consultation in person is absolutely required. Doctors would jeopardize themselves by not consulting another doctor or compromise their duty by just sending the patient's medical files. Hence, consultation over the phone "is impossible." It is "not acceptable" and it "does not happen." As for the situation in rural areas, Dupuis asserted that those who request euthanasia are mainly cancer patients (who are examined by hospital doctors), and it doesn't occur often. The requirement of independent consultation is not compromised, and if there is no independent doctor, euthanasia is not performed.

30 Leenen's letter dated February 2, 2001.

LACK OF REPORTING

Next I asked about the worrisome data on the lack of reporting. The question was formulated as the following: Record keeping and written requests of euthanasia cases have improved considerably since 1990; there are now written requests in about 60% and written record-keeping in some 85% of all cases of euthanasia. The reporting rate for euthanasia was 18% in 1990, and by 1995 it had risen to 41%. The trend is reassuring, but a situation in which less than half of all cases are reported is unacceptable from the point of view of effective control.[31] What do you think? I then added, how can the reporting rate be improved?

Most interviewees did express concern about the lack of reporting and would like to introduce changes to increase the level of reporting. John Griffiths suggests that non-criminal control on euthanasia would be more appropriate.[32] Given that doctors do not report and it is difficult to locate abuse, Griffiths thinks that a different system is needed. In his view, the only way to improve the situation is to leave the issue within the realm of the medical profession. Griffiths suggests a three-tier system: medical committees to review the euthanasia cases, plus medical inspectors, plus disciplinary committees. Instead of five regional committees, Griffiths suggests a low level of control in which each hospital would have its own review committee to examine the circumstances of death, including abstinence and pain relief. What is needed is effective control of the "whole balloon," in Griffiths' terms. Griffiths explains that pressing the balloon on one side would increase it on the other side. In other words, control of euthanasia might increase death as a result of pain relief and abstinence. Therefore, it is advisable to establish a committee in each hospital, whose role would be to review all cases of death and to refer questionable cases to medical disciplinary committees.[33]

Similarly, Bert Thijs predicts that the reporting rate will be improved if the threat of prosecution is lessened. He hopes that the introduction of the regional committees will improve the reporting rate, insofar as the committees are closer to the medical profession and have no legal authority. Whereas previously all cases went to the public prosecutor, now the committees will serve as a buffer and the role of the public prosecutor will decrease accordingly.[34] Thijs believes that another means to improve reporting is through medical education. Promoting discussion on the ending of life in medical schools and in society at large is one example of increasing social control through education and communication.

31 John Griffiths, "Effective Regulation of Euthanasia and Other Medical Behavior that Shortens Life," draft paper (October 14, 1998), pp. 10, 11; John Griffiths, Alex Bood and Heleen Weyers, *Euthanasia and Law in the Netherlands, op. cit.*, pp. 236–237.

32 In his comments on the first draft of this chapter, Griffiths wrote that he does not regard the situation as perfect, "the imperfections are a matter of concern, and something should be done about them. As a matter of fact, something is being done: unlike other countries, the Dutch are continually working on the adequacy of control of this sort of intrinsically dangerous medical behavior." Personal communication on July 10, 2000.

33 For elaborated discussion, see John Griffiths, Alex Bood and Heleen Weyers, *Euthanasia and Law in the Netherlands, op. cit.*, chap. 6.

34 Cf. Jacqueline M. Cuperus-Bosma, Gerrit van der Wal, Caspar W.N. Looman and Paul J. van der Maas, "Assessment of Physician-assisted Death by Members of the Public Prosecution in the Netherlands," *Journal of Medical Ethics*, Vol. 25 (1999): 8–15.

Some interviewees think that *the* major problem in the practice of euthanasia in the Netherlands is the low level of reporting.[35] They said that 41 percent of reporting is unacceptable. Several explanations for this finding were given: (1) euthanasia came under the Penal Code and doctors feared possible prosecution; (2) preference of secrecy, as part of the doctor-patient relationship. Physicians wished to maintain trust with their patients and felt euthanasia was a private matter; (3) laziness on the part of doctors who wished to avoid the paper work; (4) many physicians were also willing to lie at the patient's/family's request or for their own personal reasons.[36]

While trusting doctors, the view is that doctors need to report because euthanasia should never become a routine action. Euthanasia should be considered an extraordinary measure to be employed in extraordinary circumstances. Doctors should discuss their conduct in the open and expose the practice of euthanasia to public scrutiny. Several interviewees[37] emphasized that doctors need not worry if they follow the Guidelines. They think that the new proposal – to report to the regional committees – might bring some improvement. In this context it is noted that Leenen, who on the whole is skeptical about the work of the committees, nevertheless estimates that seventy percent of the Amsterdam cases were reported in the SCEA project in 1999. In turn, Jaap Visser also thinks there is an improvement in the level of reporting. However, he estimates that of the euthanasia cases only 55–60 percent is now reported.

G.F. Koerselman does not share the optimism about the positive role of the regional committees. He thinks the regional committees would not change much. He believes the organization is secondary, and it is the *value system* that is important. At this point, almost no one contests the vital policy decisions that were made. Koerselman added that even if the regional committees improve the level of reporting, a change in the climate is what is really needed.

Henk Jochemsen thinks the regional committees might improve the level of reporting, but like Koerselman he does not think this is the real issue. Physicians are now more aware of the Guidelines, there is more pressure on them to report, and we can assume this pressure will continue. Jochemsen's impression is that the committees are and will continue to be tolerant of the physicians. The committees also educate physicians on the proper performance of euthanasia, and in his view, this will help to make euthanasia even more a part of society. Similarly, Chris Rutenfrans does not see great importance in the regional committees as they receive only the politically correct cases, those performed according to the Guidelines. He thinks the level of under-reporting is quite high.

35 Sjef Gevers, Rob Houtepen, Ruud ter Meulen, Ron Berghmans, Henk Leenen and Egbert Schroten.

36 Van der Maas and his colleagues stated that after performing euthanasia and assisted suicide, three quarters of the general practitioners and about two thirds of the specialists reported "natural death" in the declaration of death. The most important reasons for falsely declaring natural death were: the "fuss" of a legal investigation (55%), fear of prosecution (25%), the desire to safeguard relatives from judicial enquiry (52%) and bad experiences in the past with stating non-natural death (12%). P.J. van der Maas, J.J.M. van Delden, and L. Pijnenborg, *Euthanasia and other Medical Decisions Concerning the End of Life*, Health Policy Monographs (Amsterdam: Elsevier, 1992), pp. 46–48. See also Gerrit van der Wal, Paul J. van der Maas, Jacqueline M. Bosma *et al.*, "Evaluation of the Notification Procedure for Physician-Assisted Death in the Netherlands," *New Eng. J. of Med.*, Vol. 335, No. 22 (November 28, 1996), esp. p. 1707; Martien Tom Muller, *Death on Request* (Amsterdam, Vrije Universiteit Thesis, 1996), p. 73.

37 Van Leeuwen, Gevers, Schroten and Wijsbek.

Van der Arend would be surprised if the regional committees manage to substantially increase the level of reporting. He thinks the committees will generate more paper work, with meager substantive positive results and will not change the GP's inclination not to report euthanasia cases.

Many of the interviewees found it necessary to call other countries into their conversation, always with the purpose of apologetically arguing that the situation in the Netherlands is no worse than in those countries. The same line of apologetic tone can be found in Dutch publications authored by scientists who fundamentally agree with the policy of euthanasia. Consider, for instance, the following concluding statement of an article compiling a very brief sketch of reports about incidence of euthanasia, assisted suicide and "actions intended to hasten a patient's death" in the Netherlands, Australia, the UK, the US, Denmark and Norway:

> ...the conclusion is that EAS [euthanasia and assisted suicide] is occurring in medical care at the end of life in all countries studied... Most worldwide surveys on incidence of EAS show lower figures than those reported in the Netherlands, where there is a lenient policy for prudent practice. Yet, in the Netherlands the actual incidence of EAS is lower than the number of requests received; more requests are refused than granted.[38]

H.J.J. Leenen, who has been instrumental in his efforts to change the Penal Code so as to permit voluntary euthanasia, said that outside the Netherlands no one reports. Doctors perform euthanasia and the act is reported as a normal, natural death. If the Dutch want to conduct euthanasia in the open, it should be adequately controlled and the reporting needs to be full and complete. The Dutch Medical Association accepted the new law proposal, which Leenen helped to formulate, stating that a physician who performs euthanasia but does not report it will be prosecuted for murder.

In a letter commenting on this study, Leenen wrote that in France in 1991 thirty-six percent of the doctors said that they had administered euthanasia once or twice a year. In the United Kingdom the percentage was twelve in 1994. In these countries the level of reporting was zero. The same is true for the United States where Leenen knows that euthanasia is administered. So, Leenen asks, "how can you blame a country in which many cases are reported and a policy exists to increase the reporting – although I agree that we have not yet reached our goal – while you say nothing about [other] countries?"[39]

Den Hartogh mentioned religion as a significant factor that might hinder reporting. In the orthodox Protestant communities, doctors are more reluctant to perform euthanasia. Some would refuse, and others would refer patients to another doctor. Those who are willing to perform it would do so secretly and would fail to report. Den Hartogh said that when the regional committees were established, he had hoped that their existence might lead to an improvement in reporting. However, this has yet to happen.

George Beusmans revealed that his experience with reporting (which involved calling a coroner) was not very good. He explained that the practice of euthanasia constitutes an intimate moment between himself and the patient's family; and when an intruder (the coroner) arrives, that intimacy is destroyed. When the patients had families, he advised them that it was *not* necessary to call a coroner. However, Beusmans maintained that during the

38 Martien T. Muller, Gerrit K. Kimsma and Gerrit van der Wal, "Euthanasia and Assisted Suicide: Facts, Figures and Fancies with Special Regard to Old Age," *Drugs & Aging*, Vol. 13, No. 3 (September 1998), at 190.

39 Letter dated February 2, 2001.

last few years he did call a coroner. Ten years ago (i.e., 1989), euthanasia was more the exception, but currently it is practiced more often, and Beusmans now has more experience with euthanasia. He and his colleagues talk about it in their continuing education programs. Interestingly, Beusmans does not think that the regional committees will make any difference.

Gerrit Kimsma admitted that he did not report his first euthanasia case. He was convinced that he was doing the right thing and that the law lacked sensitivity. He claimed that his second case took place several years later, and that he did report it. From then on, Kimsma reported all his cases. He thinks that physicians serve a social role with a professional obligation to society, and thus they are required to report. He believes that it is unprofessional not to obey the Guidelines, and that doctors should not fear prosecution if they conform. In regard to the regional committees, Kimsma is unsure about whether they will increase the level of reporting. He testifies from his experience as a member of one regional committee that of 300 to 400 cases examined, there was only one case in which a physician was deemed not careful enough in the euthanasia procedure.[40]

Bert Keizer, who like Beusmans and Kimsma practices euthanasia, provided a very candid and thorough reflection on reporting and the need to improve it during my interview with him in April 2002. Keizer testified about his own experience, saying that in 1983 he decided to euthanize one of his patients. He tried to consult with one of his colleagues but all were afraid of prosecution, so he performed the euthanasia without consultation, and without reporting. The act was done in secret, with no assessment, "a conspiracy almost," said Keizer, between him and the patient. Keizer maintained that the first ten euthanasia cases were done in secret, and the death of the ten patients was reported by him as a natural death.

I asked Keizer whether he regrets any of his euthanasia cases and he answered that he regrets one: the case of a 56 year old dying patient who had been treated by oncologists for two years, after which period they gave him up, saying that there was nothing more they could do for him. The patient was furious. He had undergone very painful and drastic treatments upon the assumption that there was hope, and after all those harsh treatments he realized that the doctors had been misleading him. The disillusioned patient went to Keizer and asked for death, and Keizer complied. Keizer did not consult anyone and, looking back, he regrets that he failed to consult a colleague, and that he assisted in this patient's "revenge on life" as it were. Keizer emphasized that of the twenty two euthanasia cases that he was involved in, he regrets only this one.

By the end of the 1980s, the atmosphere changed and Keizer felt more at ease consulting and reporting cases of euthanasia. Some court cases received public attention. The Dutch Royal Medical Association addressed the issue with a more favorable eye. The National Voluntary Euthanasia Society grew stronger. There was a more sympathetic attitude to performing euthanasia and PAS. Keizer could have found physicians to consult before performing euthanasia. In any event, said Keizer, he always observed all the rules of careful conduct, with the exception of consultation and reporting. He always asked the euthanasia

40 In his comments Kimsma wrote that the percentage of reported cases is the Achilles heel of the practice. But one must realize also that in the non-reported cases physicians and families are involved and in their opinion often the law has no place at the sickbed. Kimsma disagrees strongly with this position but he acknowledges that some physicians do not report because "they still have bad memories of reported cases in the past and have lost trust in the prosecution." Personal communication on September 9, 2000.

request in writing, and he always spoke with the patients' families. Since the late 1980s he consulted with colleagues he knew who were working in other Amsterdam nursing homes.

In 2002, anonymous surveys indicate that 50 per cent of all cases are reported. Keizer is not impressed by the argument that doctors fail to report because of the paper work involved. The paper work takes only five minutes. He thinks that there are what he terms "cowboy doctors" who do not want to be controlled. They cherish their independence and are certain that they are doing the right thing. Keizer objects to this way of thinking, arguing that there should be some mechanism of control. He first contended that reporting could possibly be improved by trying to change the psychology of doctors, but on second thought he feared that this cannot be done.

A. van Dantzig and Heleen Dupuis were sympathetic in regard the physicians' lack of reporting. They both said that physicians performing euthanasia do not want to be bothered with filling out forms and waiting months to find out whether they will be prosecuted. Van Dantzig regards the institution of the regional committees as an improvement and a preferred mechanism over the criminalizing of euthanasia. Dupuis claims that lack of reporting is the consequence of legal ambiguity. Physicians who feel that their behavior was moral do not see why they need to comply with the bureaucracy. Van Dantzig and Dupuis think that euthanasia should remain in the realm of medical practice, not of criminal law.[41]

VIEWS ON THE GUIDELINES

The critical line of questioning suggests that the Guidelines are not working; that they cannot be regarded as providing a sufficient control mechanism to prevent abuse. The answers of most interviewees affirm that the euthanasia policy and practice in the Netherlands suffer from some basic flaws. I wanted to examine the mechanism of guardedness. Most of my interviewees answered my preliminary question as to whether they are satisfied with the Dutch practice of euthanasia in the positive. After reviewing the failure of the Guidelines, I asked: Are you happy with the Guidelines?

This question came immediately after discussing the evident deficiencies in consultation and reporting. Still, many interviewees said that they are content with the Guidelines. This is quite a puzzle. Some of them[42] stated succinctly in one single sentence that they are content with the Guidelines on euthanasia: The Guidelines are good, generally enforced, and there is no need to change or amend them. Others have reservations and proposed amendments to improve the policy.

Rob Houtepen said that he is not only happy with the Guidelines, but he could not think of better ones. Heleen Dupuis asserted that the Guidelines provide workable criteria, but, of course, they cannot cover everything. In the final analysis, we have to rely on the competence and accountability of physicians who perform euthanasia. Dupuis and Schroten stated that they know the medical profession very well and, on the whole, trust the doctors involved. Most of them are very responsible, and it would be ludicrous to assume that there is some sort of a criminal gang in the medical sector whose aim is to kill their patients.

41 See also van Dantzig's general statement in *Phase II*. Personal communication on July 6, 2000.

42 H.J.J. Leenen, A. van Dantzig, Bert Thijs, Gerrit van der Wal, Henri Wijsbek, Ron Berghmans and George Beusmans.

Similarly, Johannes van Delden approves of the Guidelines, as they require an explicit request from the patient in situations without prospect for cure or recovery. If no reasonable palliative care is available to relieve suffering, then considering euthanasia is permissible.

Paul van der Maas and Sjef Gevers were ambivalent about the issue, feeling that while the Guidelines are fine overall, some amendments should still be made. Gevers has two suggestions for improvement: specify how to treat demented patients, and clarify the role of advance directives. The Guidelines say nothing about advance directives, and he feels that they should address this issue.[43] In this regard, Ron Berghmans is worried about advance directives for demented patients.[44]

Ruud ter Meulen and James Kennedy assert that more discussion is needed in Dutch society about the meaning of the terms "helplessness," "misery," "suffering," "unbearable," etc. These terms, used in the Guidelines, should be less ambivalent and more specific. They feel that these terms are open to various interpretations. Ter Meulen objects to the general trend that the Netherlands is following and accuses the country of being much too liberal and unclear about the criteria that facilitate euthanasia. He claims that some basic questions about palliative care remain unanswered, particularly about what to do if the patient rejects palliative care and insists on euthanasia. Kennedy asserts that the ambiguity on the issue is evident, especially when one looks at pertinent court rulings. There have been numerous occasions on which the Guidelines were not honored, but the physicians involved were exonerated because their intentions were perceived to be genuine and noble, and the Guidelines were interpreted in a way that supported their conduct.[45]

43 On the other hand, Ruud ter Meulen explicitly rejected the proposal (within the new law) to give advance directives or living will a legal status. Decisions for euthanasia should be based on medical criteria and enduring request of the patient. A living will, he argued, is not an enduring request. Demented patients cannot make an explicit request and should for this reason be kept out of the euthanasia practice, even if they have made up a living will. Ter Meulen maintained: "I do think that children are not able to make a real assessment of their medical situation, thus I am against the proposal to give children the right to ask for euthanasia. This has also been left out of the new proposal, after severe criticisms." Personal communication on August 23, 2000.

44 *Nederlands Tijdschrift voor Geneeskunde,* Vol. 143 (1999), at 17, reported the case of a 71-year-old man who for four years had had a psycho-organic disorder, diagnosed as deteriorating cerebral atrophy and multiple brain infarction. He asked his doctor to help him to die because he did not want to cope with further decline. The hospital's chief psychiatrist, a committee of independent professionals, and an external consultant psychiatrist assessed the patient's case. Four months after assessment the patient drank a high dose solution of barbiturate given to him by his doctor and died. The public prosecutor approved the procedure after consultation with the national forum of procurators general. The case raised fears that it might bring euthanasia for demented elderly patients a step closer. The Dutch Alzheimer Foundation warned: "Dementia itself could never be a reason for assisted suicide because the patient is incapable of making an informed request." The doctor involved in the case defended her actions, saying that she opposes euthanasia and assisted suicide for patients who are unable to express their will, and she agrees that most patients with dementia cannot. Her patient, however, was lucid at all times and completely able to understand the consequences of his request. Medical Director of the Royal Dutch Medical Association, Rob Dillmann, said that if a patient was in the early stages of dementia but still clearly competent and with an untreatable progressive neurological disease then there was the possibility of appropriate PAS. Cf. Tony Sheldon, "Euthanasia Endorsed in Dutch Patient with Dementia," *BMJ,* Vol. 319 (July 10, 1999), at 75. For further discussion, see R. Cohen-Almagor, *The Right to Die with Dignity* (Piscataway, NJ.: Rutgers University Press, 2001), chap. 5.

45 For discussions of the leading legal cases, from *Postma* to *Chabot,* and the legal changes, see *Chapter 2* and John Griffiths, Alex Bood and Heleen Weyers, *Euthanasia and Law in the Netherlands, op. cit.*, pp. 51–88. For outcomes of criminal prosecutions during 1981–1995, see *Ibid.*, p. 245. See also Carlos F. Gomez, *Regulating Death* (New York: The Free Press, 1991), pp. 25–39.

Evert van Leeuwen believes that the Guidelines should specify the drugs used for euthanasia and should be more explicit in regard to consultation requirements. The requirement for an "independent consultant" is open to interpretation, and a stricter definition is in need to specify that the consultant needs to be a truly independent expert.

John Griffiths reiterated that the system is ineffective and should be changed to rectify the inadequate reporting and difficulties of enforcement in cases of euthanasia. The Netherlands must function under a different system in which the medical profession controls itself.

Five interviewees are not happy with the existing situation, arguing that the Guidelines have not been followed.[46] Den Hartogh would like to institute effective control mechanisms to improve the level of reporting and provide adequate consultation. He also contends that many physicians do not understand the legal situation and that the euthanasia policy should be clarified to avoid misunderstanding and diffuse unnecessary fears. Van der Arend and Koerselman made two suggestions:

(1) Insist on providing quality palliative care (see *Chapter 10*).[47]

(2) Include in the Guidelines the requirement that the patient must be in the terminal stage of disease. Koerselman says that there should be a link established between euthanasia and life expectancy by limiting the practice to situations involving the final stage of an incurable disease. Such an inclusion would limit the number of euthanasia cases and would help prevent a slippery slope.[48]

In this context, I recall a conversation I had with a physician who practices euthanasia. I asked him whether he regretted any of the cases. Without hesitation or thinking, the doctor reacted by saying "No." I asked him: Please do not answer so quickly. Reflect for a minute. The doctor reiterated his conclusive "No." I pressed him for the third time, asking him to think about the patients that he had had throughout the years, ponder for a minute or two, and only then answer my question. After such a reflection, the doctor said "well, maybe one case," and told me the following story about the very first time he performed euthanasia. The case involved a man who suffered from cancer and had lost the will to continue living. He wanted to have full control over his life and death, even to determine the moment of his death. The doctor explained to him that he was not in the final stage of the disease, that he anticipated the patient still had a few more months to live, and that the euthanasia act could wait. The patient would not take "no" for an answer. He persisted with his request until the doctor could not stand his nagging and complied despite feeling that the act was premature. The doctor explained to me that the patient was "quite a character" and absolutely determined to decide his destiny, where and when he wanted his life to end. Looking back on this case years later, the doctor acknowledged that he should not have succumbed to the pressure. During those years, doctors who performed euthanasia did not consult a colleague and all was left to the patient and the doctor.

[46] Govert den Hartogh, Chris Rutenfrans, Henk Jochemsen, Arie van der Arend and G.F. Koerselman.

[47] Van Delden stressed that because euthanasia is the last resort, there is a place in the Guidelines to specify the option of providing palliative care prior to euthanasia. Jaap Visser explained that the Guidelines were the outcome of consultation between the medical profession and jurisprudence, and they are basically good. Their problems are in implementation and lack of experience. Physicians should be more aware of the possibility of palliative care and should refer patients to alternative treatment prior to opting for euthanasia.

[48] Kimsma also believes euthanasia should be available as an option only at the final stage of life.

The next chapter addresses the need for adequate palliative care and asks whether there is a "culture of death" in the Netherlands that makes euthanasia a simple, maybe too simple and premature, option.

CHAPTER 10

ON PALLIATIVE CARE AND THE DUTCH CULTURE

PALLIATIVE CARE

Herbert Hendin testified that his experience with a few physicians who had performed or been consultants in dozens of euthanasia cases indicated that they were uninvolved in palliative care.[1] Zbigniew Zylicz regarded the lack of hospice care and the fact that there were only 70 palliative care beds in the country as reflections of having the easier option of euthanasia. He argued that palliative care was virtually unknown in the Netherlands and that people mistakenly equated palliative care with the use of morphine or other drugs, not understanding that it involved much more than the use of painkillers.[2] The next question, therefore, was posed as follows: It has been argued that the policy and practice of euthanasia is the result of undeveloped palliative care. What do you think? I also mentioned the fact that there are only a few hospices in the Netherlands.

Many interviewees agreed with the statement.[3] Almost all of those agreeing with it said that only during the late 1990s were people beginning to admit that there was a need to improve palliative care. Today there are pain specialists in nursing homes and hospitals. Despite the increased investment in palliative care on the part of the government, the interviewees argued that still more attention should be paid to it, especially in medical schools and hospitals. Visser, Schroten, van Delden, Koerselman and van der Arend insist that doctors first need to explore other options for helping the patient prior to choosing euthanasia. Some interviewees said that the situation in the Netherlands on this sphere is no different than the situation in other countries.

In van der Arend's terms, there is a balance between cases of euthanasia and the quality of palliative care. If there is poor palliative care, then the number of euthanasia cases will increase, and *vice versa*. Van der Arend suggests including in the Guidelines a requirement to provide good palliative care before considering euthanasia. He maintains that the quality of palliative care in the Netherlands is not high enough and that physicians are lacking the information on how to give palliative care because it is not part of medical education. There is also a need to increase the number of hospices in the country.

1 Herbert Hendin, "Euthanasia Consultants or Facilitators?" *MJA*, Vol. 170 (1999), pp. 351–352. URL: http://www.mja.com.au/public/issues/apr19/hendin/hendin.html

2 Zbigniew Zylicz, "Euthanasia," *Lancet*, Vol. 338 (1991), p. 1150; *idem*, "Palliative Care: Dutch Hospice and Euthanasia," in David C. Thomasma, Thomasine Kimbrough-Kushner, Gerrit K. Kimsma, and Chris Ciesielski-Carlucci (eds.), *Asking to Die* (Dordrecht: Kluwer Academic Publishers, 1998), p. 196. See also H. Matthews, "Better Palliative Care Could Cut Euthanasia," *BMJ*, Vol. 317 (1998), p. 1613; Bert Broeckaert and Rien Janssens, "Palliative Care and Euthanasia: Belgian and Dutch Perspectives," *Ethical Perspectives*, Issue 9/2 (June 2002): 156–176; T. S. Jost and D. Mendelson, "A Comparative Study of the Law of Palliative Care and End-of-Life Treatment," *J. of Law, Medicine and Ethics*, Vol. 31 (2003): 130-143.

3 Arie van der Arend, Govert den Hartogh, Ruud ter Meulen, G.F. Koerselman, Henk Jochemsen, Evert van Leeuwen, Egbert Schroten, James Kennedy and Chris Rutenfrans.

The most vocal critic of euthanasia, G.F. Koerselman, contends that with adequate palliative care, euthanasia becomes unnecessary. Good palliative care enables people to continue living and coping with suffering without opting for euthanasia. Koerselman stated, "palliative care is absolutely under-developed. None of our medical faculties has a professor of palliative care."

Henk Jochemsen notes that only the Catholic University of Nijmegen has a professor for pain relief.[4] Jochemsen also sees this as an indication of the place of palliative care in the Netherlands. He argues that when the discussion on euthanasia evolved during the 1970s and 1980s, the field of palliative care was developed in many parts of the world. In the Netherlands, there was no room for both concepts to develop simultaneously and as euthanasia became an accepted practice, palliative care was pushed aside. Jochemsen maintained that the Guidelines are not precise enough, explaining that the KNMG statement depicts the performance of euthanasia only as a last resort. If we follow this statement seriously, then it means that the GP performing euthanasia needs to consult a palliative care expert prior to the mercy killing. However, GPs are not equipped to decide on the various alternatives designed to alleviate suffering. Furthermore, if the patient is suffering mentally, then the GP has to consult a psychiatrist. Jochemsen thinks that the government and the KNMG have recently put more emphasis on palliative care.

Ruud ter Meulen argues that the criticisms are correct: Euthanasia is practiced too quickly because palliative care is not a real option. The acceptance of euthanasia has led to the undermining of palliative care. Hospitals, though somewhat more advanced in this sphere than nursing homes, still have a long way to go. In comparison, Belgium, for instance, boasts a far better practice of palliative care.

Evert van Leeuwen and Egbert Schroten explained that until the 1980s, it was argued that specialists in palliative care were not needed. Consequently, new developments in the field were not adopted, and the issue was left quite neglected. During the 1990s, the idea of using palliative care experts became increasingly accepted, but there remains much to do in this area. James Kennedy, an historian who studies the origins of the euthanasia policy, clarified that there are very few hospices in the Netherlands because home care is the prevalent phenomenon and the role of the GP in providing care is conceived to be sufficient. It should be noted that van der Maas contests this view. He wrote in his comments that the low number of hospices in the Netherlands is attributable not to the specific role of the GP, but to the fact that many nursing homes have specialized departments for end-of-life care.[5]

Likewise, Ron Berghmans argued that palliative care is well developed in nursing homes, where care for spiritual and psychological needs is provided in addition to care for physical needs. In contrast, palliative care is less developed in the practice of individual GPs. Berghmans did not believe that palliative care would eliminate euthanasia. At most, it might decrease the number of requests. He also doubted whether the policy and practice of euthanasia had actually frustrated the development of palliative care. Berghmans confirmed that recently the government had stimulated the development of palliative care by making available large amounts of money for universities to develop units and expertise in the field.

4 In his comments, Arko Oderwald wrote that this is not true. The Free University of Amsterdam also has such a professor, and maybe there are more pain relief professors. Oderwald added that there are certainly more specialized pain centers in university clinics and other hospitals. Personal communication on August 28, 2000.

5 Personal communication on September 18, 2000.

It is intended that these trained experts will provide consultation to GPs who deal with terminal patients.

Johannes van Delden said that the fact that the Netherlands does not have many hospices should not be taken as an indication that "we don't have palliative care." Rather, "we organize things differently."[6] Van Delden asserted that doctors know how to deal with pain. GPs do receive adequate education and training in this sphere, and, in any event, palliative care involves more than just ameliorating pain. It also involves providing psychological support for the patient. While acknowledging that there is room for improvement in this area, van Delden argued that euthanasia has actually paved the way for calling more attention to palliative care. He emphasized that euthanasia does not exclude palliative care, but the reverse, and adamantly objected to the accusation that lack of palliative care has resulted in greater acceptance of euthanasia. Van Delden noted the problem that sometimes patients refuse to receive palliative care and expressed his uncertainty about whether doctors should comply with euthanasia requests under such circumstances.

George Beusmans substantiated van Delden's arguments, testifying that he received medical education on palliative care and contending that this aspect does not present a major problem. He considered himself to be qualified in evaluating suffering and in providing proper medication and palliative care. However, pain is not the primary issue in question. People that request euthanasia suffer not only from physical pain, but also from anguish, dependence on others, and anxiety regarding the unknown. They ask for help because they have lost the sense of purpose in waiting and see no good reason for prolonging their situation.

Bert Keizer is in agreement with Beusmans and disagrees with Koerselman's statement about the reverse connection between adequate palliative care and the number of euthanasia requests. Keizer testifies that in twenty years of practice he treated between 900–1000 dying patients. He has a lot of experience, also with providing palliative care. Interestingly, Keizer himself did not take any palliative care courses. He does not think he needs such courses. Keizer believes the practice of palliative care in Dutch nursing homes is fine, and that the quality of palliation in hospitals is improving. Keizer objects to the thought that improving palliative care may reduce the number of euthanasia requests. He explains that people who initiate euthanasia requests are desperate. They ask for death not because they are in pain, but because they lost hope. Keizer attributes the improvement in palliative care to the international critique. Now there are advanced courses on palliative care, the issue is on students' curricula, medical staff discusses the limits of medicine.

Govert den Hartogh provided the most extensive answer, claiming that the flip side of the fact that GPs are involved in euthanasia is that they may not be fully aware of all palliative care options themselves and hence have to rely on expert advice and assistance. While acknowledging that palliative care cannot serve as a complete substitute for euthanasia, den Hartogh insists that euthanasia should be used only as the last resort. The doctor should first try to alleviate suffering by other means and resort to euthanasia only when all those other means have failed. At the same time, den Hartogh does not think that Dutch doctors are less knowledgeable about palliative care than doctors in other countries. Furthermore, den

[6] Ms. Els Borst-Eilers gave a similar explanation. Cf. Proceedings of *Euthanasia and Assisted Suicide in the Netherlands and in Europe*, Maastricht, June 10 11, 1994 (Luxembourg: Office for Official Publications of the European Communities, 1996), pp. 64–65.

Hartogh holds that while palliative care is available to patients in hospitals, many patients still choose to leave the hospital and return home to die by requesting euthanasia. This can be seen as an indication that euthanasia is in fact performed on many occasions after palliative care has been tried.

In his comments on the first draft of this study, den Hartogh elaborated on his answer, arguing that it is wrong to think that palliative care can provide a viable alternative to euthanasia in all cases. There are many instances, in which no pain relief can be given, except by sedating the patient into unconsciousness. Furthermore, we need to take into account the many other forms of suffering besides pain, which can be part of the process of dying, and for which relief may be unavailable or insufficient. Den Hartogh rejects the suggestion that palliative care is neglected because the Netherlands has euthanasia as an alternative. It is essential in this respect to see palliative care as an integral part of medical care for the dying, whether it is given in an institutional setting or (as the Dutch generally prefer) at home. It is nowadays a normal practice for a GP who is responsible for the treatment of cancer patients to act in constant consultation with a hospital's pain-relief team. Of course, there is more to palliative care than pain relief.[7]

It seems to me that the developed practice of euthanasia came at the expense of developing adequate palliative care. Only during the last few years, the Netherlands has started to invest in this sphere. This is a very positive development. Many times patients may contemplate ending their lives because they are unable to cope with their pain. Once good palliative care is developed we may expect that the number of euthanasia requests will be reduced; this is provided that general practitioners will offer this option first and not euthanasia, and that they are able to provide good palliative care to their patients, or at least refer them to palliative care experts. The entire euthanasia practice could be transformed for the better once nursing-home doctors, specialists, and especially general practitioners will be aware of the positive consequences of good palliative care at the end of life.

CULTURE OF DEATH

The next question was framed as follows: Daniel Callahan argues that there is a "culture of death" in the Netherlands.[8] What do you think? I intentionally refrained from explaining the term "culture of death." I wanted to see whether the interviewees have different ideas on what would constitute such a culture.

This question came late in the interviews. Sometimes I did not ask it for fear that the interviewee might be offended. Indeed, the question did upset some of the interviewees.

The majority of interviewees rejected Callahan's statement *tout court.* A small minority that objects to the euthanasia policy and practice agreed with the statement. Henk Leenen was surprised by my question. He said: "We do not have a 'culture of death.' We strive to prevent the slippery slope. In the 1960s, the same criticism was voiced in regard to the legalization of abortion. Presently, the Netherlands, with its liberal policy on abortion, has the lowest rate of abortion in the world, in proportion to its size. There was no slippery slope as a

[7] Personal communication on August 27, 2000. For further discussion, see R.J.P.A. Janssens and H.A.M.J. ten Have, "The Concept of Palliative Care in the Netherlands," *Palliative Medicine*, Vol. 15 (November 2001): 481–486.

[8] Discussions with Dan Callahan at the Hastings Center, New York (January–February 1994, March 1999).

result of the liberal abortion law." Leenen is certain that the Netherlands does not have more euthanasia cases than other countries. The problem is that there is no way to prove this because little amount of data is available in regard to what is taking place in other parts of the world.[9] Leenen concluded his answer by saying that Callahan is speculating without any hard data. On a personal level, it seems to Leenen that Callahan does not trust human intentions. After all, most people have respect for other people, and most physicians are not criminals. Physicians act in a *bona fide* manner to fulfill the wishes of their patients.

Rob Houtepen argued that there is no general climate of death and that there are no shortcuts to performing euthanasia, as Callahan alleges. He acknowledges that there is a need to be on guard, but "there is no room for alarm." At the present time, Houtepen admits, the Guidelines are insufficiently followed, but the 1990 and 1995 reports do not indicate a slippery slope.

Interestingly, later in the interview, Houtepen suggested that hospital physicians thrive on action. In their quest to cure, they always seek something else to do, something to explore. This is their culture. Consequently, euthanasia is not a popular practice in hospitals, and there is no need to worry about physicians' conduct. "Relatively speaking," Houtepen elucidated, "the weak point are the GPs." In hospitals and nursing homes, there are mechanisms of social

[9] An estimated 37,000 euthanasia deaths occur each year in Australia as a result of doctors intentionally accelerating a patient's death, according to a confidential new survey of 3,000 Australian doctors. The survey defined euthanasia as any death "intentionally accelerated by a doctor." It included deaths caused by a doctor withholding or withdrawing treatment. This is quite a broad definition, well beyond the contemporary Dutch narrow definition of euthanasia. The survey suggests that almost a third of deaths occur after doctors intentionally hasten their patients' deaths. It found that the rate of doctor-assisted death in Australia was double that of the Netherlands, and that Australian doctors were far less likely to discuss their decision to hasten a patient's death with the target patients or even seek their consent. Incidents of non-voluntary euthanasia were five times higher in Australia than in the Netherlands. Cf. Helga Kuhse, Peter Singer, Peter Baume, Malcolm Clark and Maurice Rickard, "End-of-life Decisions in Australian Medical Practice," *Medical J. of Australia*, Vol. 166 (1997): 191–196. See also Gary Spink, "Euthanasia More Common in Australia than Netherlands," *Montage*, Vol. 8, No. 1 (April 1997). I have mentioned another study that compared attitudes and practices concerning end-of-life decisions between physicians in Oregon and in the Netherlands, showing that American physicians found euthanasia acceptable less frequently than the Dutch and have been involved in these practices less frequently than the Dutch. Cf. Dick L. Willems, Elisabeth R. Daniels, Gerrit van der Wal *et al*, "Attitudes and Practices Concerning the End of Life: A Comparison Between Physicians from the United States and from the Netherlands," *Arch Intern Med.*, Vol. 160 (2000): 63–68. A third study showed that substantial proportion of physicians in the United States in the specialties surveyed reported that they received requests for PAS and euthanasia, and about 6 percent have complied with such requests at least once. Cf. Diane E. Meier, Carol-Ann Emmons, Sylvan Wallenstein, Timothy Quill, R. Sean Morrison and Christine K. Cassel, "A National Survey of Physician-Assisted Suicide and Euthanasia in the United States," *New Eng. J. of Med.*, Vol. 338, No. 17 (April 23, 1998): 1193–1201. In Belgium, surveys conducted in 1985, 1987 and 1988 showed that 34%, 40% and 52% of physicians in the respective studies admitted practicing euthanasia. Proceedings of *Euthanasia and Assisted Suicide in the Netherlands and in Europe*, Maastricht, June 10–11, 1994 (Luxembourg: Office for Official Publications of the European Communities, 1996), pp. 23–24. A recent study showed that end-of-life decisions (ELDs) are prominent in medical practice in Flanders. The frequency of deaths preceded by an ELD is similar to that in the Netherlands. Cf. Luc Deliens, Freddy Mortier, Johan Bilsen, Marc Cosyns, Robert Vander Stichele, Johan Vanoverloop and Koen Ingels, "End-of-life Decisions in Medical Practice in Flanders, Belgium: A Nationwide Survey", *Lancet*, Vol. 356 (November 25, 2000): 1806–11. In Denmark, a 1996 study among physicians showed that 30% of respondents said that they had received a request for euthanasia and that 5% had participated in the practice. In Norway, a 1997 study among physicians showed that 6% of respondents indicated that they had performed actions intended to hasten a patient's death. Cf. Martien T. Muller, Gerrit K. Kimsma and Gerrit van der Wal, "Euthanasia and Assisted Suicide: Facts, Figures and Fancies with Special Regard to Old Age," *Drugs & Aging*, Vol. 13, No. 3 (September 1998), p. 188.

control. The nursing staff works in teams, and there are usually people around, whereas GPs act more or less alone. It is worth noting that, as the data show, most of the euthanasia cases are performed by GPs.[10]

In his answer, Ron Berghmans countered the "culture of death" notion by pointing out that there are many cases in which euthanasia was requested but not performed.[11] In general, the spirit of medicine and health care is to help people and not to kill them. Berghmans explained that the Dutch climate is different from other countries insofar as patients feel that they have the right to request a quick death, to determine the moment of death, and to receive full compliance by their doctors. Consequently, patients can put a lot of pressure on doctors to comply with euthanasia requests. There is no insistence on continuing life by all means, and the emphasis is put on the meaning and quality of life. Furthermore, there are open discussions about what constitutes useful treatment. Euthanasia, PAS and painkillers are all considered legitimate mechanisms to prevent unnecessary suffering.

A. van Dantzig and Egbert Schroten dismissed the question. Schroten said that the notion of a culture of death is an exaggeration, claiming that it is much more difficult to control passive euthanasia. Sometimes doctors switch off machines sooner than required, "but this happens all over the world." Schroten added that he is not convinced that the way in which "we treat euthanasia makes passive euthanasia a greater problem. Our open discussions sharpened the awareness on the subject. We are more alert." Thus, it is "ridiculous" to speak of "culture of death."

Heleen Dupuis was puzzled and surprised by the question and like Schroten dismissed it as "ridiculous." Dupuis said succinctly and conclusively: "We don't have a 'culture of death.'" She reiterated that "'Angels of death' never happened in Holland. They happen in countries that do not discuss euthanasia in the open, like Austria, but not here." Dupuis maintained that people normally want to live. Elderly people that are suffering are the ones who usually make requests for euthanasia. She asserted that she is much more afraid of a society that denies the euthanasia option, and emphasized that doctors do not perform euthanasia on patients if they are presently unable to express their will, even if the request for euthanasia was previously made in a living will. Doctors might cease treatment, but they will not perform euthanasia under these circumstances. The patient needs to be competent in order to exercise euthanasia as an option.

[10] P.J. van der Maas, J.J.M. van Delden, and L. Pijnenborg, *Euthanasia and other Medical Decisions Concerning the End of Life*, Health Policy Monographs (Amsterdam: Elsevier, 1992), p. 40; Paul J. van der Maas, Gerrit van der Wal, Ilinka Haverkate *et al*, "Euthanasia, Physician-Assisted Suicide, and Other Medical Practices Involving the End of Life in the Netherlands, 1990–1995," *New Eng. J. Med.*, Vol. 335, No. 22 (November 28, 1996), pp. 1701–1702; B.D. Onwuteaka-Philipsen, A. van der Heide, D. Koper, I. Keij-Deerenberg, J.A.C. Rietjens, M.L. Rurup, A.M. Vrakking, J.J. Georges, M.T. Muller, G. van der Wal, P.J. van der Maas, "Euthanasia and Other End-of-life Decisions in the Netherlands in 1990, 1995, and 2001," *Lancet*, Vol. 362 (August 2, 2003): 395–399.

[11] About 37% of all serious and persistent requests will lead to euthanasia and PAS. Nearly half of the requests are refused because alternatives are still available, the request was not thoroughly considered, the patient does not have a proper understanding of the disease, or the physician has objections. Cf. Paul van der Maas and Linda L. Emanuel, "Factual Findings," in L.L. Emanuel (ed.), *Regulating How We Die* (Cambridge, Mass.: Harvard University Press, 1998), p. 158. Note that van der Maas and Emanuel entitle those requests "serious," apparently not cognizant of the inherent contradiction between the term "serious" and the reasons for not honoring those requests.

Bert Keizer laughed when he heard the question. Then he said that the term "culture of death" is awfully vague and that, on the contrary, the Netherlands has a culture of life. This term, "culture of life", Keizer explicates, is as vague as saying that "we have a culture of death." "What kind of crap is this?," Keizer asks, maintaining that 99.9 per cent of Dutch people have health insurance. This is not the case in the United States, where 40 percent of the people are not insured at all. Their situation is appalling. They have the worse hospitals, the worse treatments. This culture Keizer calls "a culture of death" or "a culture of gross neglect." Implicitly Keizer was saying that Daniel Callahan should criticize his own culture before addressing the Dutch culture.

By contrast, Koerselman, Jochemsen and Rutenfrans, complained about the atmosphere surrounding the policy and practice of euthanasia, and voiced their dissent against the institutional mechanisms that are used to de-legitimize them and undermine their position. Koerselman said that advocates of euthanasia dismissed him as a Catholic fundamentalist. In fact, he is not a Catholic at all. He was brought up without any religious background, but his critics find it difficult to believe that a secular person would object to euthanasia with so much passion.[12] He also testified that he often felt treated like a clown. The media invited him to debates on euthanasia issues only because they needed to depict "the other side," not because they were really interested in exploring the anti-euthanasia arguments. Koerselman declared that he was fed up with this treatment and with the dismissive attitude that he received from scholars and colleagues.

Koerselman is worried about the general atmosphere in the Netherlands. He agrees with Callahan that a "culture of death" has developed, with autonomy and the prevention of suffering as the prime concerns. He feels that the Netherlands is becoming a narcissistic society, a place only for happy and healthy people. And if one is not happy and healthy, one has a right to die and to have a nice funeral, with the doctor being obliged to help. Koerselman does not know where this approach will lead. The debate, he argues, started with the assumption that it is possible to shorten life when the patient is suffering and life expectancy is very short. Hence, euthanasia would only shorten life by a few hours or days. The next stage broadened the framework so that the time one is expected to live is not important; only the suffering is important. Then the discussion evolved to speaking not only of physical suffering, but also of mental suffering. Koerselman asks rhetorically: What constitutes suffering? Living with a serious handicap entails suffering. Does it mean that all handicapped patients should be put to death? Parents with children who suffer from Down's syndrome are asked: Why did you allow this to happen? How can you see your child living like this?

According to Koerselman, living wills for euthanasia that are signed by competent patients who become demented are now legally accepted.[13] The next stage might be killing these patients without satisfying the requirement of a written document. A guardian would be appointed for such patients, who would say that "this is a case of unbearable loss of dignity" and that "euthanasia is the answer." The principle of avoiding suffering thus overrides the principle of autonomy. The climate assumes that these patients are suffering and would probably opt to die if they were able to express their will.

[12] Rutenfrans noted that in the Netherlands, people with a religious commitment, how liberal this religious commitment may be, "are seen as completely out of their mind." Personal communication on August 22, 2001.

[13] Cf. "Euthanasiewens demente geldig," *de Volkskrant* (July 16, 1999), p. 1. See also p. 7.

One important factor in generating this culture, in Koerselman's opinion, is the Dutch Voluntary Euthanasia Society, which has more than 100,000 members. He thinks that this strong movement is very successful in making euthanasia an integral part of the Dutch medical establishment. The Society places considerable emphasis on the concept of "dignity," suggesting to people that they prepare a document stating that if their dignity were to be lost, they would prefer not to be treated. Then Koerselman added the following chilling statement: "If I'll be involved in a car accident and will be brought to hospital, I am not sure that all will be done to save my life. The climate is in this direction."

In my 2002 interview Koerselman mentioned an organization named *Einder* (Horizon) whose aim is to help people commit suicide. *Einder* provides detailed instructions and assistance to needy people. The Euthanasia Society recently wished to join *Einder* in this regard, and to train lay people so they could assist patients who wish to die.[14]

The concept that justifies physicians' decisions in such cases is quality of life.[15] James Kennedy explains that there is a "collective consensus" about what constitutes quality of life. When the quality of a given life deteriorates, it is unclear whether treatment will be adequately provided until death. Treatment might be stopped prematurely at an early stage, when the patient is conceived to be leading a "pointless life" or when he "suffers a great deal." In such cases, passive euthanasia and double effect are the mechanisms used for ceasing treatment. Sometimes it is euthanasia.

Marina Dackman, a nurse who worked in an oncology department at a hospital in Amstelveen, told me the story of a patient who suffered severe pains and arrived for extensive medical checkups. The examinations confirmed that the patient suffered from progressive and incurable bone cancer. The doctors revealed this information to the man who, for the first time, was diagnosed as a cancer patient. Obviously he was very upset. He was full of anxieties, in addition to pain, and was digesting this piece of information without much help from family or friends. The patient, in his mid forties, had a family but the relationships with them were not good. His wife and children were clearly estranged from him and did not offer much support. On the following morning after the discovery he initiated a talk with the doctors, telling them that he was afraid and confused, and that he would not like to suffer and to continue living in such a condition. The medical stuff was quick to react. The same evening the doctors ordered to administer him with considerable doses of medication cocktail consisted of morphine, insulin and potassium designed to kill him. Within a further 30 hours the patient was dead. Instead of providing counselling, comforting him, relieving his anxieties and fears the doctors opted for the easy solution of termination of life. The nurse thought that the patient had a few months to live, and that with adequate palliative care he could have continued living. She thought the doctors' conduct was both unethical and illegal and refused to be the one who administered the injections. This incident was a catalyst for her to leave the hospital after a few months. She lost her trust in and appreciation for the medical system.

[14] For information about Einder, contact De.Einder@freeler.nl; http://www.kuleuven.ac.be/thomas/actualiteit/indekijker/19/impulsen7.htm

[15] For discussion on the quality of life concept in bioethics, see R. Cohen-Almagor and Merav Shmueli, "Can Life Be Evaluated? The Jewish Halachic Approach vs. the Quality of Life Approach in Medical Ethics: A Critical View", *Theoretical Medicine and Bioethics*, Vol. 21, No. 2 (August 2000): 117–137.

Chris Rutenfrans expressed concern that the Netherlands allows doctors to be above the law in killing other people. This is very dangerous, especially when the necessary control mechanisms are lacking. He testified that he is not principally opposed to euthanasia, but rather to the simple generalization and legitimization of euthanasia, expanding the acknowledgement of its necessity in a few cases to many more unjustified cases.

Rutenfrans said that 95 per cent of all newspapers and magazines in the Netherlands are in favor of euthanasia; only some very small religious papers are against it. Nationwide, seven newspapers and many regional papers support the euthanasia campaign. Together, 30–40 papers express a positive stance on the issue. Most newspapers reported the positive points covered in the 1990 and 1995 reports. The problem is that they hide the negative points, and the court cases are reported with a bias towards physicians who practice euthanasia. There is no sincere attempt to tackle the issue fairly, showing both sides of the debate. Rutenfrans maintained that it is bad for one's reputation to be against euthanasia because it gives the appearance of being conservative, and it is not good to be conservative in the Netherlands. This is why Rutenfrans wants to disassociate himself from the subject and is inclined to write on other issues. In his view, the country is not very liberal, but rather is conformist in its liberalism. Its people do not want to hear ideas that clash with their liberal values.

Rutenfrans recounted that in 1986 he had co-authored a booklet against euthanasia, entitled *May the Doctor Kill,* with Caterina I. Dessaur, a novelist known under the pen name Andreas Burnier. This was a highly polemic, very controversial book. Rutenfrans maintained that Dessaur had been quite a famous novelist before publishing this book. After publication, Dessaur's consecutive novels were criticized more harshly than before, depicting her as a conservative reactionary. In effect, she was cast out of the country's literary circles.[16]

Govert den Hartogh does not agree with Rutenfrans on many issues, including this one. In his comments, he wrote that Dessaur's literary star had fallen in popularity simply because her later books were not as good as her early ones; her position on euthanasia had nothing to do with it. This is confirmed by the excessive popularity of the author Willem Jan Otten, who is as fervently opposed to euthanasia as Dessaur ever was.[17] Otten joined Koerselman and others in writing a pamphlet on the *Chabot* case, which was widely reported, discussed and also acclaimed in the Dutch press.[18]

Henk Jochemsen indicated that during the past 20 years, the general atmosphere has been in favor of euthanasia.[19] The mentality now is to stop treatment at an early stage when the patient is suffering. Quality of life has become the major principle at the expense of respect for life. Jochemsen claimed that physicians had told him about the difficulties they would face in finding a job in some institutions if they declared themselves to be opposed to euthanasia. The establishment view is pro-euthanasia, and one might be harmed if one takes a contrary view.

[16] For further deliberation, see Herbert Hendin, *Seduced by Death, op. cit.,* pp. 105–107.

[17] Koerselman wrote in his comments that the status of Otten as a writer and a poet ""has steeply fallen down since he publicly confessed having become a Catholic." Personal communication on February 20, 2002.

[18] Personal communication on August 27, 2000.

[19] A poll in 1996 showed that 84% of the population is in favor of euthanasia if a fellow human being is in an unacceptable and futureless situation. Cf. http://www.ves.org.uk

I asked Evert van Leeuwen if he thinks it would be possible to elect an anti-euthanasia professor to the Chair in medical ethics in major universities. His candid answer was "probably not" because the Chair serves as a consultant in euthanasia cases referred to him or her by hospitals affiliated with the respective universities. If it is known that the professor objects to euthanasia, then there would be no point in consulting with him or her on this issue at a time when euthanasia does take place in hospitals. Hence, it is necessary to fill important posts with like-minded people who will maintain the positive climate towards euthanasia.

In his comments on the first draft of this study, van Leeuwen noted that Henk ten Have, who opposes the practice of euthanasia, chairs the department of Ethics, Philosophy and History of Medicine in the Catholic University of Nijmegen.[20] In this university, it would indeed be highly surprising to appoint a supporter of euthanasia for that position.

In his counter arguments to the text, Govert den Hartogh made the same observation, doubting whether a euthanasia supporter would be welcome in Nijmegen. The Free University in Amsterdam is a religious institute, so it traditionally makes some requirements concerning people's opinions in selecting personnel. But the medical director of the academic hospital of the University of Amsterdam is a member of an orthodox Protestant church, and never made a secret of his opposition to euthanasia under any conditions.[21]

[20] Personal communication on August 30, 2000.

[21] Govert den Hartogh's personal communication on August 27, 2000.

CHAPTER 11

ON LEGISLATION AND THE *CHABOT* CASE

LEGISLATION

At the time of the interviews in 1999, the Netherlands appeared to be on the verge of legislating euthanasia. Indeed, sixteen months later the Dutch Lower House of parliament voted in favor of enacting the euthanasia law. I wondered at the time what the interviewees thought about the legalization motion. Thus, my next question was: Would it be preferable to legislate euthanasia? Why yes or why not?[1]

The interviewees exhibited split views on the issue. Some were in favor of legislation for instrumental and symbolic reasons. Others utilized different instrumental and symbolic reasons to argue against legislation. Three interviewees preferred to wait for some years before changing the law. Among the arguments for legislation were the following: explicit law would clarify what doctors should and should not do and would eliminate misunderstanding. It would exert effective public control on physicians' conduct and, at the same time, provide physicians with legal support. Legislation might also improve the reporting rate. In addition, interviewees said that the ambiguous legal situation diminished respect for the law and that public opinion supported legislation.

Those opposing legislation but who were still supportive of the euthanasia policy thought that the Guidelines were working just fine. They did not think that formal societal control through law was the answer because physicians disliked legal intervention, and they disagreed that the level of reporting would increase due to the legal change. Euthanasia should remain in the realm of medical practice. The minority opposed to euthanasia throughout the interviews was obviously against conferring legitimacy and legal status to this conduct thinking that legalization would move the Netherlands further in the wrong direction.

H.J.J. Leenen was against the existing ambiguous situation. He had pushed for the legalization of euthanasia for many years, and it was he who had drafted the new law. Leenen argued that euthanasia could no longer be regarded as an exception in the sense of Article 40 of the Penal Code. More than 3000 euthanasia cases in a year did not constitute an "exception." He saw the need for an explicit law to address and clarify what doctors should and should not do. Sjef Gevers also preferred legislation. He did not want euthanasia to be governed by medical practice. He believed that euthanasia decisions should be separated from other medical decisions at the end of life. Likewise, Bert Thijs and Gerrit van der Wal thought that it was preferable to legalize euthanasia so as to exert public control on doctors' conduct. Van der Wal was hoping that the Netherlands would then have the lowest rate of euthanasia in the world.

1 The law had passed after *Phase I* of the interviews. Still, it is interesting to note what leading authorities thought of the pending law.

Rob Houtepen was also in favor of explicit legislation because it would provide legal support for physicians to take what they believed was the proper action. By enacting explicit legislation, the Dutch would be expressing the view that self-chosen death is optional. Houtepen noted that legislation would also have a symbolic function insofar as euthanasia would cease to be a criminal act.

Henri Wijsbek explained that there was and is Article 293, which forbids the taking of the life of anyone at that person's explicit request; but in jurisprudence a justification for euthanasia based on Article 40 (which is also part of the formal law) was accepted by the Supreme Court since the mid eighties. This situation created confusion. Since most people believed that euthanasia was formally forbidden but tolerated nonetheless, explicit recognition of its lawfulness in a formal statute is preferable for the sake of clarity.[2]

Govert den Hartogh was in favor of legislation because the existing situation was unclear for both physicians and patients. He claimed that even Dutch medical ethicists did not understand the situation properly, often drawing an analogy between the Netherlands' treatment of soft drug use and its treatment of euthanasia, without realizing the significant difference between the two. Indeed, euthanasia was permissible if done according to the Guidelines and *was not* a criminal offence if carried out properly. Drug use, on the other hand, *was* a criminal offence, although a tolerable one. Den Hartogh thought that the new law should codify the existing situation, nothing more and nothing less. Doctors did not understand the legal situation, and an explicit law would clarify matters and eliminate further misunderstanding. At the same time, den Hartogh reiterated the need for continued social control. He speculated that legalization might contribute to improved reporting, but might also cause doctors to think that euthanasia is no longer a crime and is considered a "normal" medical practice, thus not worth reporting.

Like den Hartogh, Ron Berghmans expressed his desire to change the existing situation and to have the law clarify the conditions under which euthanasia is acceptable. Berghmans was struggling with the issue, feeling confident that effective external control is necessary but unsure as to how willing doctors would be to cooperate with external bodies.

In contrast, George Beusmans and Gerrit Kimsma were quite happy with the ambiguous Dutch policy on euthanasia. Kimsma contended that the ambivalence reflected the theoretical and practical paradoxes and provided mediation between proponents and opponents of euthanasia.[3] Beusmans argued that doctors knew about the Guidelines and that they were working just fine.

[2] Later in his comments, Wijsbek wrote that he prefers that euthanasia should fall under the same ruling as other medical interventions, and hence under the medical exception. The attitude to the violation of the practice depends on which part of the Guidelines a doctor does not abide by, whether he should be charged with murder or not. If he has only neglected to consult a colleague, he should not be charged with murder; if he kills a patient who is not suffering unbearably and has not requested euthanasia, he should be charged with murder. Comments made on February 11, 2002.

[3] In his comments, Kimsma clarified that he was not "quite happy" with the ambiguous Dutch policy on euthanasia. However, he considered it a policy that for the time being needed no change, because of its reflection of the theoretical and practical paradoxes. These paradoxes are: the need to self-report a crime, the lack of clarity on whether euthanasia is a medical or a legal act, or both; the stressing of autonomy as a condition, as expressed in the euthanasia request, while showing compassion in alleviating pain and hastening death in patients without adequate communication is a medical-ethical value that works in the mind of physicians but is not recognized as an important value. Personal communication on September 9, 2000.

Johannes van der Delden and Arie van der Arend thought that euthanasia should remain in the Penal Code as a matter of strategy and that prosecution should continue to be a background consideration for any case of euthanasia. They believed that euthanasia should remain an exception in the medical setting. Van der Delden recognized that the existing situation might be unfair to physicians. If the system condones the practice under certain circumstances, then why not make it legal? On the other hand, keeping euthanasia in the Penal Code conveys a clear message to physicians that it is an exceptional practice. Because van der Delden did not conceive of legalization as a solution, and doctors do not like lawyers "looking over their shoulders," he was ultimately in favor of leaving the situation as is. Both van der Delden and van der Arend were unsure about whether the new law would change anything in actual practice.

Likewise, Frank Koerselman argued that euthanasia should remain in the Penal Code. He believed that the Guidelines were much too tolerant and hence would not like them to become part of the law. As a doctor, Koerselman said, he did not want to become obligated professionally or morally to commit euthanasia.

Henk Jochemsen and Chris Rutenfrans felt it was absolutely necessary to prohibit euthanasia under the Penal Code. Jochemsen explained that, at the time, the burden was on the physician to prove that he had fulfilled the requirements. If euthanasia were to become permissible by law, then the burden would shift to the prosecutor to prove that the physician had breached the Guidelines. This would be very difficult to prove, given that the physician provides all the data. He believed that changing the existing legal circumstances would make euthanasia a more accepted practice and would move the Netherlands further in the wrong direction.

Egbert Schroten's opinion was quite grim. He maintained that there would always be doctors who fail to report. He believed that the public wants to have the option of euthanasia; hence, the practice would continue even if the Netherlands were to ban it – it would simply go underground. If the law were to change, then the fear of prosecution may be reduced and the level of reporting might increase. Still, Schroten was not in favor of changing the law. While maintaining that euthanasia should be forbidden in principle, he granted that exceptions should be allowed under certain conditions. He was not familiar with the new law proposal prepared by Leenen.[4]

Van Dantzig held that euthanasia should remain in the realm of medical practice, not criminal law. Unlike Keizer, he believed physicians "find it a pain" to fill out forms and go through procedures. He believed that the regional committees represented an improvement over the Penal Code and viewed the new law as a technique to address an existing practical problem. While euthanasia could be authorized by the law, this was considered to be a secondary issue for van Dantzig, as well as for Dupuis, who felt it was more important to maintain a tolerant attitude toward euthanasia in order to address a pressing moral issue. In his comments on the first draft of this study, van Dantzig wished to clarify that the forms were not what constituted the main objection, but having to endure months of uncertainty before a doctor knew whether or not he would be prosecuted.[5]

4 Cf. H.J.J. Leenen, "Bill on Euthanasia and Assisting Suicide in the Netherlands," *European J. of Health Law*, Vol. 5 (1998): 299-324.

5 Personal communication on July 6, 2000.

Three of the interviewees[6] suggest that there was no need to rush the process and recommended waiting for a period of time before writing a new law (Kimsma said ten years; van Leeuwen two years), while evaluating the work conducted by the newly established regional committees. After collecting more information about the process, the Netherlands could then decide how to pursue the matter, keeping the data in mind. Van Leeuwen warned against politicizing the issue. He felt that politicians were exploiting it for their partisan interests, cynically presenting themselves as proponents of humanistic values, while in practice lacking the necessary knowledge to actually draft a new law. Kimsma explained that such a law would not only reflect the situation but would also influence the situation, which in the Netherlands had already gone far enough. In turn, Jaap Visser thought that euthanasia should remain in the Penal Code to remind physicians that they must follow the Guidelines for careful conduct or else they would be prosecuted.[7]

My own feeling was that the situation at the time of the interview was problematic and unhealthy. It had resulted from the natural Dutch inclination to reach a compromise. But you cannot have your cake and eat it too. Evidently, creating such grey legal areas was baffling and confusing to all parties. The appeal to *force majeure* had been developed in criminal law for exceptional situations, but euthanasia was conducted hundreds of times each year on a regular basis. Furthermore, the situation did injustice to both the physician and the patient. Although physicians should not be afraid to report cases of euthanasia if they followed all the Guidelines, the reality showed that given the confusing situation, many physicians were still unsure of themselves and opted for the safer way of not reporting. Therefore, my feeling was that the confusion should have been cleared and the Guidelines should have been anchored in law. The law should be very detailed, explicitly stating the conditions for the performance of voluntary euthanasia and physician-assisted suicide (see *Conclusions* of this study).

CHABOT

A question related to the former question on legislation was: What do you think of the *Chabot* case?

The *Chabot* precedent, described in *PART A,* has generated debates and heated discussions in the Netherlands as well as in other parts of the world. It concerns an issue that was rarely discussed up to that point: physician-assisted suicide and euthanasia in psychiatric practice. Subsequent to the Supreme Court ruling, the Dutch Ministers of Health and Justice commissioned research to study the phenomenon. In March 1996, a one-page questionnaire was sent to a sample of 673 psychiatrists. Of the 552 respondents, 205 (37%) had at least once received an explicit and persistent request for physician-assisted suicide from a patient. Twelve (2%) had at least once assisted in suicide. An additional 345 respondents (64%) thought physician-assisted suicide for psychiatric patients could be acceptable; of those, 241 said they could conceive of a situation in which they would be willing to assist in suicide.

6 Jaap Visser, Gerrit Kimsma and Evert van Leeuwen.
7 For a critical discussion of the law, see I.G. Finlay and B. van Dijk, "Euthanasia: The Dutch Experience and What It Entails in Practice," *Lancet Oncology*, Vol. 3 (March 2002): 135–136.

Detailed information was obtained from 202 respondents about their most recent requests, with 43 of them (21%) reporting that they contemplated granting the patients' requests for PAS. Of those, 40 consulted one or more colleagues each.[8]

Nearly all respondents said that one or more psychiatrists should be consulted on such a matter. The main reasons for consultation were to assess whether the phenomena of transference and counter-transference (the patient's unconscious feelings and attitudes toward the therapist and *vice versa*) might have influenced the decision-making process (50%), whether the request was well-considered (58%), and whether there were still remaining treatment options (58%). Of 537 respondents, 438 (82%) thought that the psychiatrist consultant should always examine the patient; 93 (17%) thought this was necessary in some but not all cases, and 6 (1%) thought it unnecessary.[9]

Let me proceed by providing an account of my communications with Dr. Chabot. As said before, he was the only person who explicitly declined my invitation for an interview. Chabot referred me to three writings on his court case,[10] saying that Hendin's account of his case was not authorized by him, and in fact gave a tendentious account of his discussion of the case with him.[11] Chabot ended his personal communication by saying that neither before the Boomsma case in 1991 nor since the Supreme Court verdict in 1994 "have I ever assisted someone again in his/her suicide. Of course, many depressed patients have asked my help to commit suicide. But contrary to the impression some have given about my professional stance I know how to give adequate treatment for that request."[12]

After a careful reading of the suggested writings I wrote to Dr. Chabot again, asking him to consider seven queries pertinent to his case as summarized in *Chapter 2*. Below are presented word for word the questions and the answers.[13]

Q1. Is it a common practice within psychiatrists' circles to rely on transcripts as a substitute to meeting with patients?

A1. Yes, it definitely is. In fact it is very exceptional for a consultant psychiatrist to see a patient himself after reading transcripts of sessions. A reason for doing so might be that the transcripts don't make sense or that he/she has good reason for not trusting the content of the transcripts. In previous verdicts by the Dutch Supreme Court, no obligation for a consulted doctor to see the patient himself had ever been formulated before my case. I can certainly understand the requirement now and I wish one of the consulted physicians had asked me to

8 Johanna H. Groenewoud, Paul J. van der Maas, Gerrit van der Wal *et al*, "Physician-assisted Death in Psychiatric Practice in the Netherlands," *New Eng. J. of Med.*., Vol. 336, No. 25 (June 19, 1997): 1796–1800.

9 *Ibid.*

10 J. Griffiths, A. Bood and H. Weyers, *Euthanasia and Law in the Netherlands* (Amsterdam: Amsterdam University Press, 1998); B. Sneiderman and M. Verhoef, "Patient Autonomy and the Defence of Medical Necessity: Five Dutch Euthanasia Cases," *Alberta Law Review*, Vol. XXXIV, No. 2 (1996); A.J. Klotzko, "Arlene Judith Klotzko and dr. Boudewijn Chabot Discuss Assisted Suicide in the Absence of Somatic Illness," *Cambridge Quarterly of Healthcare Ethics*, Vol. 4 (1995): 239–249.

11 Chabot maintained: "It may be useful for you to know that four other interviewee's by dr. Hendin in Holland (prof. van der Maas, prof. Schudel, dr. Cohen and dr. Dillmann) have objected in writing to the editor of that Journal [*Issues in Law and Medicine*] to express their criticism of the skewed presentation of their words by dr. Hendin." Personal communication on June 5, 1999.

12 Personal communication on June 5, 1999.

13 Personal communication on August 8, 1999.

see my patient. But there was no way to know any obligation existed to do so at the time, in 1991.

Q2. Had you asked the lone dissenter, who thought that you should persist in treating Mrs. Hilly Bosscher, to meet with her? If yes, what was his reaction? Did they meet? If not, why didn't you ask the dissenter to meet with her?
A2. No. He was definite in his view and showed no interest in either testing or strengthening his view by seeing my patient.

Q3. At the stage when you met with Dr. V., had you already decided to comply with Mrs. Bosscher's request?
A3. Yes.

Q4. Why didn't you ask Dr. V. to examine Mrs. Bosscher?
A4. Dr. V. was not consulted by me to give his agreement with my decision. The suggestion in Dr. Sneiderman's paper that Dr. V. was consulted by me to test once more my views on Mrs. B. is not correct. Yes, I did discuss her wish to die with him but more as a good friend as well as experienced family doctor than in a formal professional way as I had done with the others. I decided to ask him to join me to be present as a professional observer who could possibly testify (if need be) later in court whether or not I behaved in a professional and responsible way. Don't forget that I had never dealt before with physician assisted suicide. What if, against all odds, she failed to die? I felt more comfortable to have a doctor-witness present apart from the witness-friend of Mrs. B.

Q5. Dr. V. attended Mrs. Bosscher's home on the day of her death. Was it the first time that he met her? Why then?
A5. Yes, it was the very first time. He knew me as a conscientious psychiatrist and he accepted my need for a doctor-witness. So he didn't feel any need to check on my decision. He was definitely not present in the role of a consulted doctor.

Q6. How many doctors, who met with Mrs. Bosscher, shared your view that her situation was hopeless?
A6. Only her own family doctor met her between May 1991 and her death. He opposed assisted suicide on principle (being of a Christian denomination). She knew this and therefore she did not seriously enter discussions with him about her motivation to die. This family doctor was later questioned about his views on her by the Medical Disciplinary Board. He expressed himself critical of her death wish saying that "in his experience after heavy weather there always was bound to follow sunshine." But the Board did not use his views in their judgment. I guess this was so because they felt his negative views on Mrs. B.'s death wish were more determined by his general worldview than by professional considerations.

Q7. Was there the possibility for pressing the option of anti-depressant therapy more vigorously?
A7. I don't know how I could have pressed that option further given that she said (after stressing the importance of trying them several times): "Even if antidepressants will make

life more bearable I still want to die because by mourning and by growing over my losses I will become a different person than I was when I was a mother and happy. I don't want to become that different person. It would feel for me like becoming disloyal to my two sons." I (and so did others later) think this is a well-considered refusal: loyalty to her sons was the prime value of her life. By being treated and forgetting about them she felt she would become disloyal to her 'core-self.' I urged her to enter mourning therapy, preferably with antidepressants, arguing that after the war people came back from the German extermination camps with less than she had still now and that some of them had grown over their losses and meant a great deal to the next generation; so why couldn't she become a 'different person' and help others overcome their losses? She thought it over and then refused saying: "that is them, I am a different person. My life is over. If you continue insisting on treatment it is better our ways part now and I will not bother you further with my request." It was then that I stopped pressing for antidepressants.

As expected, the interviewees exhibited contradictory views on the *Chabot* case. Some supported what Dr. Chabot did and consequently objected to the Supreme Court decision.[14] Others had mixed views about the case.[15] The majority of interviewees thought that Dr. Chabot had acted unprofessionally and in haste and therefore considered the court decision to be proper and justified.[16] That six experts condoned what Chabot did and saw no room for alarm will no doubt evoke concern among euthanasia critics.

The two psychiatrists, A. van Dantzig and Frank Koerselman, exhibited the two polar views. Their professional assessments of the case coincide with their general views on euthanasia. I will outline the different views on a scale from the whole-hearted support to the passionate opposition. The strongest supporter of Chabot among the interviewees was van Dantzig, who said that euthanasia prevents people from dying alone. This was the first time that he had been consulted about a euthanasia case. Van Dantzig was certain, as was his colleague Chabot, that Mrs. Bosscher would have committed suicide anyway. She was not going to accept any further treatment or therapy. Chabot had the choice of helping her to die with her relatives present or to let her commit suicide alone. Mrs. Bosscher refused to take antidepressants. Van Dantzig did not think she was depressed; it was a natural reaction to the death of her two sons. If van Dantzig had felt that pills could help, he would have ordered that Mrs. Bosscher be institutionalized and forced to take the pills. But this was not the case here. There was no cure for her. Suffering from depression is not different from suffering caused by cancer. Suffering is suffering. Van Dantzig maintained that in the current environment, it is needed that another expert sees the patient. At that time, however, this was not the case.

In his comments on the first draft of this essay, van Dantzig added that he did not see the patient because he was asked by Chabot to examine the procedure and to assess if it was complete. That van Dantzig did find to be the case. Chabot had correctly diagnosed his patient as suffering unbearable pain from losing both her sons. His patient was firmly resolved to try to commit suicide again if euthanasia was denied (she had made a serious

14 A. van Dantzig, Heleen Dupuis, Henri Wijsbek, Henk Leenen, Govert den Hartogh and Rob Houtepen.

15 John Griffiths, J.K. Gevers, Ron Berghmans and Arie van der Arend.

16 Gerrit van der Wal, Jaap Visser, Egbert Schroten, Evert van Leeuwen, Gerrit Kimsma, Johannes van Delden, Bert Keizer, Ruud ter Meulen, Chris Rutenfrans, Henk Jochemsen and Frank Koerselman.

suicide attempt before, but failed, and now wanted to be sure of dying). For these reasons, van Dantzig could concur with Chabot's conclusion and still does. If Chabot had come to him now with this request, van Dantzig would want to see the patient. In that respect, concluded van Dantzig, times have changed.[17]

Heleen Dupuis agreed with Chabot that Mrs. Bosscher qualified for assisted suicide. She was not depressed in the psychiatric way that may be conducive to treatment. Dupuis reiterated that people should be allowed to choose the moment of dying, maintaining that if people wanted to die, "what's wrong with providing them the pills?" Doctors are the only people who have access to those drugs. It is far more humane than to force patients to jump from bridges or buildings. If people feel that death is a good solution, why not help them? Dupuis acknowledged that most people, however, do not share her view. They feel that it is not morally acceptable to provide the pills. Doctors are reluctant to help incompetent patients to die because they are unsure of the will of the patient in the present condition. They would be willing to withhold or withdraw treatment. Morally speaking, Dupuis thinks that end-of-life decisions are about tolerance. Dying is a private matter, and it should be kept as a private choice. Society should provide safeguard mechanisms, but should allow euthanasia and PAS as options. Dupuis seemed to have a very good experience with doctors, trusting them whole-heartedly without reservations.

Henri Wijsbek published a prize-winning essay in which he defended Chabot's decision to assist in the suicide of Mrs. Bosscher on the grounds that she had a right to protect the narrative unity or authenticity of her life. He called it unusual that the Supreme Court convicted Chabot and did not refer the case to a lower court. As for the facts of the case, Chabot complied with all existing rules. Mrs. Bosscher did try other forms of therapy when her first son committed suicide, but it did not help her. It is impossible to impose treatment against the patient's will. Her suffering was unbearable, and she could not go on living. Furthermore, Chabot asked one of his colleagues to see her, but he said that there was no point, having learned everything he wanted to know from Chabot's transcripts.

H.J.J Leenen considered it unclear as to whether the patient's psychological suffering was a reason for euthanasia. Mrs. Bosscher was not a psychiatric patient, according to the accepted creteria. On the one hand, Leenen thinks that psychiatric patients should be seen by a consultant. On the other hand, Leenen felt a lot of sympathy with Chabot insofar as the patient was suffering and could not be cured. Only one of the seven consultants indicated that there was potential therapy for her. As a result of this case, the Supreme Court introduced the new requirement that a psychiatrist needs to consult another expert, who must see the patient before euthanasia or assisted suicide is performed. However, this requirement did not exist at that time. Like Wijsbek he wondered: "how could you blame someone for failing to do something that was not required at that time?" For this reason, Leenen thinks that the decision of the disciplinary court to reprimand Chabot was unfair. The court was too hard on him, and Leenen feels that it was not a fair trial. The court ignored that the majority of consultants sided with Chabot.

Govert den Hartogh seemed to be influenced by Wijsbek's reasoning. He argued that Chabot acted in a very careful way. He did not act upon an impulse, but rather thought about what he was doing. Maybe he should have invested more time, but the disciplinary court wrongly condemned him. Mrs. Bosscher was competent to make her choice. There was no

[17] Personal communication on July 6, 2000. For further discussion condoning Chabot, see Peter Singer, "Voluntary Euthanasia: A Utilitarian Perspective," *Bioethics*, Vol. 17, Nos. 5-6 (2003):533-535.

real available form of treatment. Chabot tried to find an alternative solution to help her and concluded that no treatment was available. At the same time, den Hartogh admitted that he was unsure about whether a doctor should have cooperated with Mrs. Bosscher to kill her, and that Chabot should have insisted on a consultant seeing her. This aspect is important in order to maintain control.

The response of Rob Houtepen was fascinating in its contradictions. He seemed to condone what Chabot did although recognizing that his conduct was faulty. Houtepen said that there were two problems in the case: not securing adequate consultation and not exploring alternatives for treatment. At the same time, Houtepen said that there was "no reason" to worry about the case. It might be defensible to assist in the suicide of patients in depression, who have a history of depression. Chabot's conduct was not defensible, but if the procedures are properly followed, then euthanasia should be considered an option for depressed patients. Houtepen apparently assumes that after the *Chabot* precedent, all of the intricate questions concerning depressed people will be resolved and psychiatrists will be bound to follow the procedures.

Four interviewees had mixed feelings about the case. John Griffiths seemed the least concerned among the four about the case, probably because he could not agree with the Supreme Court verdict on a matter of law. Griffiths said that he disagreed with some aspects of what Chabot did, in particular in relation to consultation. He thinks that, in retrospect, Chabot himself would agree with this. However, like Wijsbek and Leenen, Griffiths thinks it highly objectionable to apply what in effect was a new rule – consultation in person – retroactively.[18]

Gevers acknowledged that suffering might be terrible for non-somatic patients. At the same time, he agreed with the Supreme Court decision that there should be consultation in person and that the therapy should not include discussions on euthanasia. Non-somatic patients do not have the freedom to reject medical alternatives. It was not enough that Mrs. Bosscher rejected therapy to make Chabot opt for euthanasia. In turn, Berghmans and van der Arend held that there are some psychiatric cases in which euthanasia or PAS can be accepted. Both practices should be considered as options only after exhausting all other alternatives. In this case, it is unclear whether Mrs. Bosscher could have found meaning in her life. Van der Arend said, like his colleague Houtepen, that the *Chabot* case is an extreme case that almost surely will never happen again. He agreed with the Supreme Court on the legal point (which Wijsbek, Leenen and Griffiths think was flawed), but ethically he was in agreement with Chabot. It is almost inescapable to help such a patient with euthanasia. Independent consultants, however, should see the patient and should evaluate the communication between the patient and the doctor.

18 See Griffiths' analysis in "Assisted Suicide in the Netherlands: The *Chabot* Case," *Modern L. Rev.*, Vol. 58 (March 1995): 239–248.

In his comments on the first draft of this study, Berghmans wrote that a central concern in *Chabot* and other similar cases is the availability of possible alternatives to PAS. In Mrs. Bosscher's case, the issue was whether or not she might have found meaning in continuing her life after having tried antidepressant medication. While Berghmans certainly would not deem coercive treatment acceptable, he thinks that a refusal of treatment may block PAS. However, he would not automatically jump to the conclusion that Chabot ought to have denied PAS. Much depends on the particular circumstances, the aspects of a specific treatment option (e.g., the prospects, what kind of "benefits" and what kinds of risks and burdens, the time frame, and earlier experiences of the patient).[19]

The third group of interviewees objected to Chabot's conduct mainly, but not only, because of lack of consultation in person. They also mentioned Chabot's unprofessional conduct, his failure to exhaust all alternatives for treatment, and his inability to resist Mrs. Bosscher's request.

Chris Rutenfrans and Henk Jochemsen clearly objected to Chabot's conduct. Rutenfrans is worried that the Supreme Court opened the door in *Chabot* for psychiatrists to euthanize their patients. Although Chabot was found guilty, the court acknowledged that upon satisfying certain requirements, it is possible to provide euthanasia to psychiatric patients. Rutenfrans also disliked the conduct of the media in their coverage of the case. The newspapers gave a very supportive view of Chabot, at first even depicting him as a hero. Later, when the Supreme Court found him guilty, Chabot ceased to be a hero, but the media still cast a positive eye on the crossing of another border – allowing the performance of euthanasia in psychiatric cases.

Henk Jochemsen contended that even if you accept the need for euthanasia on some occasions (which Jochemsen does not), Chabot did not try enough, did not explore all possible alternatives for treatment, and behaved unprofessionally by becoming too close to his patient. Jochemsen further noted that very few psychiatrists would present themselves to the Dutch Voluntary Euthanasia Society as people who would be willing to consider euthanasia, as Chabot did. This was quite unusual.[20] After the case was published, Chabot became a hero in the pro-euthanasia circles.

Gerrit van der Wal, Jaap Visser, Egbert Schroten, and Evert van Leeuwen did not side with Chabot for the same reasons that Houtepen mentioned. Unlike Houtepen, they seemed more worried about the case. They thought that a consultant should see the patient, and that in this case there were other treatment options for the patient, which Chabot should have insisted on exploring. Schroten said that he respected the autonomy of the patient, but how can we speak of autonomy in times of depression? Schroten said that Chabot should have refused the euthanasia request, and that if Mrs. Bosscher did not accept his decision, she was always free to go to another psychiatrist. Van Leeuwen said that there are psychiatric reasons for euthanasia, but not in this case.

Three of the physicians, Gerrit Kimsma, Johannes van Delden and Bert Keizer also objected to Chabot's conduct. Kimsma explained that all of his euthanasia patients were terminal. He would not do what Chabot had done. Moreover, like van Leeuwen, Kimsma thought that Mrs. Bosscher was not a patient in the psychiatric sense. She was shopping

19 Personal communication on August 22, 2000.
20 Cf. Herbert Hendin, *Seduced by Death, op. cit.*, p. 61.

around for someone to help her. She was a client, not a patient. Furthermore, there was the development of a friendship beyond the doctor-patient relationship and problems of transference and counter-transference. Koerselman, *infra,* elaborated on this issue. Similarly, van Delden and Keizer concurred that they would not have acted in the way Chabot did. Van Delden criticized Chabot for relying too heavily on the autonomy factor, indicating that there must also be present *unbearable* suffering. It is not the role of the doctor to assist in the suicide of his patients when life gets tough. In turn, Keizer doubted whether he himself would have complied with the patient's request. He would not wish this to be his responsibility. Keizer would not have provided the patient with the lethal medication because he would be reluctant to take responsibility for her life, and death. The patient made Chabot feel that her problem is his problem. But it is not for the doctor to decide whether she should continue with her life. The patient was physically able and he does not see her mental state as a sufficient reason to cease living. Generally speaking, Keizer thinks psychiatric patients constitute a different category and should not be viewed as somatic patients. He can understand euthanasia requests initiated by somatic patients and would see such requests by psychiatric patients as justified only when they are also terminal patients.

Ruud ter Meulen noted that the *Chabot* case sheds further light on the difficulties involved in the understanding of the concept of suffering. While it was possible to reach an agreement on euthanasia for cancer patients, it is far more problematic to agree on euthanasia in relieving psychological suffering. Ter Meulen thought that Chabot was rightly condemned. Like Kimsma, he felt that Chabot was too involved in his relations with Mrs. Bosscher and failed to maintain sufficient distance.[21]

The most vehement opposition to Chabot's conduct was expressed, unsurprisingly, by G.F. Koerselman. Evidently, he was deeply disturbed by the conduct of his colleague, and rightly so. He was worried when he first heard about the case and became even more so after reading Chabot's book[22] about his patient. It appeared that Chabot was unable to diagnose Mrs. Bosscher. She suffered from depression, something that is quite normal in her condition. Koerselman's view was that Mrs. Bosscher also had a personality disorder. She focused totally on her sons. There were pathological relations between her and the older son, as evidenced by her involvement in the son's relationship with his girlfriend.

Like Kimsma, Koerselman expressed a compelling argument about the counter-transference problem that existed in this case. Chabot had personal feelings for Mrs. Bosscher. Yet, he did not realize his growing emotional involvement in the treatment. Many doctors try to relieve guilt feelings that stem from their personal past in their treatment of patients. The problem in this case was that Mrs. Bosscher made *everyone* feel guilty and was very aggressive. Many people, including Chabot, felt guilty for their inability to help her. Furthermore, the entire process took only two months. This is an extremely short time for therapy.

21 One of the interviewees who wished to be withdrawn from the book was "extremely shocked" by Chabot's conduct. Evidently, Mrs. Bosscher was unhappy. It was a pity that the Supreme Court ruled that she suffered from a psychiatric disorder. Chabot himself said that she was not a psychiatric patient. Doctors should not provide euthanasia to unhappy people. They should restrict their conduct to helping medical or psychiatric patients, not to solving unhappiness. For this interviewee, this was a step taken too far. If euthanasia should be allowed for psychiatric patients, then the procedure should be extra careful and not be conducted in Chabot's way.

22 Boudewijn Chabot, *Zelf Beschikt (Chosen Faith)* (Amsterdam: Balans, 1993).

Koerselman regarded Chabot's conduct in the case as extremely unprofessional. He explained that psychiatrists should not talk for hours with a patient in one session so as not to become over-involved. Chabot asked Mrs. Bosscher to take a room in a hotel near his work place because he wanted to spend more time with her. Subsequently, Chabot even allowed her to sleep in his house.[23] In his book, he appeared even to be proud of it.[24] He lost all professional guidelines. Like Jochemsen and Rutenfrans, Koerselman was upset by the fact that Chabot became a hero for helping a woman in distress. Her autonomous right to end her life was denied by her GP and others, and he was applauded for freeing her from her miserable life.

Koerselman did see one positive aspect in Chabot's behavior: that he talked about the case in the open. Koerselman testified that he was unaware of what was accepted conduct in psychiatrist circles prior to that time. He said: "I don't know what was the common practice. I, myself, would always want to see the patient. I was amazed that leading psychiatrists were willing to comment on the patient without seeing her in person."[25]

I asked Koerselman whether the Amsterdam medical tribunal's penalty of *berisping,* or reprimand, was serious. His answer was that receiving a reprimand is a rather serious punishment. It is more serious than a warning. Next on the scale is a payment penalty, and the most serious penalty is revoking of the doctor's license. The latter has been done only rarely, when, for instance, a doctor sleeps with his patient. In the Netherlands, Koerselman said cynically, "you can kill your patient but you should not sleep with her."[26]

[23] Chabot categorically denies this allegation, saying this is "obvious slander," completely untrue and as such well known to others, most relevant to the police and the judges that investigated his case. Personal communication on July 17, 2000.

[24] Chabot regards this statement as another example of slander: "Anybody who can read Dutch can confirm that there is no such sentence [expressing pride] in any of my writings." Personal communication on July 17, 2000.

[25] For further deliberation, see Herbert Hendin, *Seduced by Death, op. cit.,* pp. 101–102.

[26] Chabot didn't like this statement. In his remarks he wrote that this quote of Koerselman "discredits to any serious reader not so much the *Chabot* case (that was investigated by so many courts) but rather your presentation of the discussion in the Netherlands." Personal communication on July 17, 2000.

PHASE II: INTERVIEWEES' GENERAL COMMENTS

A. PRELIMINARIES

As previously mentioned, I sent the first draft of this study to the interviewees in July 2000, kindly requesting them to submit their comments by September. Most of the interviewees complied with my request. The following discussion outlines their general comments on the study and my reactions to the interviewees' critique. Their specific comments were delineated in the previous analysis.

B. GENERAL COMMENTS

Rob Houtepen was quick to respond. Houtepen states that there is a broad paradigm difference between my thinking and his. Simply put, so Houtepen claims, he is conferred with the Dutch trait of pragmatism. He explains:

> This does not preclude moral debate, but it does imply a negative heuristic concerning the absolute nature of such debate and a positive heuristic concerning the subject [euthanasia], scope and procedure of such debate. In my view most Anglo Saxon ethics, and indeed Anglo-Saxon (specifically American) culture as a whole, is characterized by a dichotomous right-or-wrong approach to moral issues. I would like to see moral deliberation performed in a practical rather than foundational mode. To me, the word practical also refers to the Aristotelian tradition of virtue ethics: the moral art is to strike the right balance between competing claims, whilst acknowledging that such a balance may differ from person to person and from situation to situation. Against this background, it may no longer be as illogical as you make it appear, to acknowledge that there are many unresolved issues and to contend at the same time that there is no need for great worry.[1]

Shortly afterwards Professor van Dantzig sent me a detailed response. His specific comments were incorporated into the text. He ended his communication with the following sketch of his views on euthanasia:

> I think that the primary task of a physician is to alleviate or end suffering. If this is not possible but by death, this option should be reluctantly taken – as amputating a leg is. When a person does not suffer, death should not be considered… A request for euthanasia I consider to be a proposition of the patient for a certain treatment for a problem. A doctor would never blindly follow such a proposal in other cases, for example an operation, and should not do so in cases of euthanasia. He should actively seek other, less extreme solutions, and my experience is that that is often a possibility. I do not think that euthanasia should be given by one doctor without consultation, and I am in favor of regulating this consultation so that all necessary expertise has been consulted. But, and it is a great but, this should be taken out of criminal law. This obligatory consultation should not be installed to prevent abuse, but to enhance quality. This would be a good thing for other important medical decisions also, and in this way euthanasia would have the benefit of lagging behind as an acknowledged medical procedure. In short, make euthanasia part of the medical practice of the highest quality.

1 Personal communication on July 3, 2000.

Van Dantzig thought that I have too gloomy a picture of the situation in the Netherlands. He testified that he had never seen cases of euthanasia without consent. In his opinion, regulations will not prevent crime: "We trust doctors not to remove our breasts or legs without very good reason, and the same should go for euthanasia." In order to ensure that their reasons are good, maintained van Dantzig, we install professional control, thereby setting an example for the rest of medicine.

Van Dantzig felt that these views are not correctly represented in this study. He explained that he had gone into this matter so thoroughly because "I deeply appreciate your effort, am convinced of your good faith, and want to help you to make the result as good as possible." He suggested that I should not look at euthanasia as if the main problem is preventing abuse. Instead, the main problem is assuring good quality of practice – "as we would want for the whole of medicine."[2]

Govert den Hartogh provided the most extensive comments, some of which were reported *supra.* Den Hartogh wrote that at several points, I suggested the existence of a slippery slope. Indeed, it seems to me that the Dutch have shifted the emphasis from autonomy to mercy, from requiring a competent request to mercy killings without such a request. In den Hartogh's view, this is turning things upside down. Euthanasia on newborns happens about fifteen times a year; on psychiatric patients five times a year, with half of those cases involving severe somatic illnesses as well; on PVS patients never or almost never (because the easy alternative is to limit care and let them die); on Alzheimer patients also never. There is no evidence of any substantial increase in any of these forms of life-ending decisions. The worrisome Remmelink data about the 1000 patients, as discussed, constitute a completely different matter, though the evidence does not point to any increase here either.

Den Hartogh added that recent evidence seems to prove that giving higher doses of morphine, even much higher ones, to patients who already are using some morphine, does not shorten their lives; it may actually rather have the opposite effect. This means on the one hand that in most cases in which doctors report giving morphine with the intention of shortening the patient's life, their action does not have the intended effect. It also means that of the "1000 of Remmelink" in up to 70% of the cases the death of the patient was not really the result of the doctor's intervention.

If there is anything worrisome about the Dutch practice, argued den Hartogh, it is rather that many patients believe in their right to euthanasia on request and that many doctors do not firmly insist on being personally convinced of unbearable suffering for which there is no alternative. So what we need in order to prevent "slippery slopes" is to combat the idea that an autonomous request is enough; hence, a strong insistence on the principle of mercy. To illustrate this further: Euthanasia in PVS and dementia cases cannot be justified by mercy at all, but may be justified by respect for autonomy, if we take advance directives to represent the will of the patient. This is the main point of discussion concerning the new law.[3]

For my part, I am not altogether convinced that the Dutch have a fully informed perspective on "mercy killing" performed on newborns and babies. Den Hartogh says that there are 15 euthanasia cases of newborns each year. He has no qualms using the term "euthanasia" in this context, although surely the newborns did not give their consent. Den Hartogh also does not discuss babies. Because many births take place at home, and this issue

2 Personal communication on July 6, 2000.
3 Personal communication on August 27, 2000.

may be considered as a private matter between medical professionals and families, it might be the case that the scope of killing is much larger. Not enough research has been conducted on premature death of babies, especially those who were afflicted by diseases like Down Syndrome.

Considering the overall atmosphere in the Netherlands of condoning "euthanasia," even when not all the conditions for the practice are fulfilled, and with the emphasis on relieving suffering (of the patient and sometimes, implicitly, of his/her surroundings), there is room to think that patients might not receive all the care that they deserve, and on occasion their lives might be shortened by an active measure. The fact that all of the Guidelines have been breached and yet very few amendments have been made in recent years, and that the majority of the Dutch leaders in the field accept the practice and defend it despite all the pitfalls is worrying.

As for the issue of advance directives: if patients make advance directives in the form of a living will, DNR order, a letter, etc., that they wish to continue living, no matter what, and we have no reason to believe that they have had a change of mind, then we should respect that wish.[4]

There may be cases when a patient has prepared an advance directive explicitly stating that all treatment be terminated upon reaching the last stage of an incurable disease, but the patient's present wishes cannot be fully ascertained. This may be the case, for instance, if the patient is in Post-Coma Unawareness, a condition that is termed PVS, and the attending physicians have diagnosed the condition as irreversible. In such an event, we should respect the advance directives and let the patient die. For such persons who have prepared advance directives, asking to die upon reaching a certain condition, death is not seen as the worst possible situation when compared to being on the verge of death and then being stabilized without hope of ever really getting better.

Den Hartogh maintained that basically, with the recent exception of Belgium, in no other country in the world but the Netherlands has there been any attempt to assemble the actual facts about medical decisions at the end of life. Referring to my scrutiny of the unwelcoming Dutch attitude toward critique, "Until Israel has a Van der Maas-report, I feel you hardly have any standing to make this kind of reproach."[5]

I admit that this argument puzzles me for several reasons. The Dutch tend to justify the practice in their country by comparing it to other countries. Many euthanasia activists say that the Netherlands is doing well in comparison to other countries. However, this logic avoids the main issue. The argument that the situation in other parts of the world is worse doesn't justify wrongdoing in their own country. Patients who are put to death without their consent (or, more accurately, the consent of their loved ones) will find little encouragement, if any, knowing that a similar phenomenon takes place in other countries.

Israel, like most countries around the world, does not allow the practice of euthanasia. Many Dutch scientists suggest that physicians in many countries are secretly doing what

4 For further deliberation, see Lawrence J. Schneiderman and Nancy S. Jecker, *Wrong Medicine* (Baltimore and London: Johns Hopkins University Press, 1995): 63–64; Editorial, "How Living Wills Can Help Doctors and Patients Talk about Dying," *BMJ*, Vol. 320 (June 17, 2000): 1618–1619.

5 Personal communication on August 27, 2000.

Dutch physicians are doing openly.[6] However, this suggestion is dubious. There are not enough data to either support or refute this suggestion.

Furthermore, the van der Maas reports lack constructive criticism of their findings. When asked why the first report did not include recommendations, for instance, that doctors discuss their plans with competent patients, van der Maas answered that it was understood that the report was not a policy document and hence would contain as few recommendations as possible. This was necessary to secure and retain the cooperation of the KNMG and the participating doctors.[7] Hendin rightly commented that in the interest of maintaining harmony, virtually all of the explanations of the physicians appear to have been accepted at face value, even when follow-up questions seemed necessary.[8] Critical analysis is clearly missing.

Moreover, the apologetic tone runs through the many writings of van der Maas and van der Wal. In one study, van der Maas and colleagues seemed to justify the 1000 cases of involuntary euthanasia.[9] They said that almost all cases involved patients with only a few hours or days left to live. Referring to a well-known article in *JAMA,* in which the anonymous author described the mercy killing of one of his/her patients, Debbie,[10] van der Maas and colleagues contended that the case could be defined as LAWER (life-terminating act without the explicit request of the patient). They do qualify this statement as follows: "One important difference between this case and most cases in our study is that 'Debbie's' physician saw her for the first time on the day that her life was cut short, while in our study the physician had usually known his patient for months or even years."[11] The study to which they refer was sponsored (like many of van der Maas' and van der Wal's studies) by the Dutch Ministry of Welfare, Health and Cultural Affairs and the Ministry of Justice. One must ask to what extent scientists are free to voice their opinions on intricate practices when their research is directly funded by the government that is responsible for these practices. This is an open, much debated, question.

Paul van der Maas obviously noticed my critical tone of science sponsored by the state. In his reflections on the first draft, he wrote:

> I consider myself an independent researcher, with a primary responsibility in collecting reliable data and basing impartial estimates and interpretations on that empirical information. I see no position for myself in a pro versus contra euthanasia debate and I think such a debate is entirely unproductive. As a researcher I think my responsibility is to find out what people do and how that might fit in high quality end of life medicine. During the last years part of our study has been replicated in Australia and Belgium and we have obtained funding from the European Union for an international collaborative study in order to establish empirical comparisons between countries.[12]

Den Hartogh wrote that his own attitude concerning opponents of euthanasia is the following:

6 L. Pijnenborg, "The Dutch Euthanasia Debate in International Perspective," in *End-of-Life Decisions in Dutch Medical Practice* (Rotterdam, 1995), Thesis, esp. at 119–131.

7 Herbert Hendin, *Seduced by Death, op. cit.*, p. 77.

8 *Ibid, ibid.*

9 Loes Pijnenborg, Paul J. van der Maas, J.J.M. van Delden and Caspar W.N. Looman, "Life-terminating Acts without Explicit Request of Patient," *Lancet,* Vol. 341 (May 8, 1993): 1196–1199, esp. at 1198.

10 Anonymous, "Its Over Debbie: A Piece of My Mind," *JAMA,* Vol. 259, No. 2 (January 8, 1988), p. 272.

11 Loes Pijnenborg, Paul J. van der Maas, J.J.M. van Delden and Caspar W.N. Looman, "Life-terminating Acts without Explicit Request of Patient," *op. cit.*, p. 1198.

12 Personal communication on September 18, 2000.

> I fully respect the opponents (in the Netherlands, for example, the theologians Beemer and Rothuizen; in the USA philosophers like Arras and Bok) who are prepared to grant that there is at least a problem of severe suffering at the end of life. (Similarly the proponents should take seriously the possibilities of abuse and concede that there is no possible form of regulation that has no undesirable side effects.) I welcome any opportunity to argue with people like Koerselman who, however one-sided in their views, make points that are worthwhile of consideration (e.g. the dangers of 'collusion' between patient and physician). But I don't feel obligated to acknowledge any form of opposition. So I will never make any reference to people like Rutenfrans or Fenigsen who have publicly compared Dutch euthanasia to the Nazi-practice, and who present 'data' that they know to be untrue.[13] If they feel ostracized, they have themselves to blame. And if you think this is a form of censure by silence, I plead guilty. I was puzzled how you could make such a statement, while your own reporting shows such a great diversity of opinion. The explanation may be that the spectrum of the positions taken in the Dutch discussion is very different from the spectrum of positions taken in the USA or in Israel, so the plurality of voices is not the plurality you are used to. But why should it be? My own position is, let's say, right-of-center in the Netherlands and would be pretty radical in Germany, and in the Dutch discussion Koerselman and Jochemsen represent minority views while in most other European countries they would be standard ones. We quarrel, often in un-Dutch, almost Israeli polemical style, about mercy and autonomy, about the precise specification of the guidelines, about assisting suicides of psychiatric patients, euthanasia of Alzheimer patients, whether children should have the option of requesting euthanasia, etc. But not even Koerselman opposes euthanasia under all circumstances. The focus of our discussion is no longer legalization or not, because a full and effective legal prohibition is by now not an open political option at all. If only for that reason, it is nonsense to talk about a "euthanasia campaign" in the press: if any such campaign ever took place, it has been won years ago. As a matter of fact almost everything that is published these days is critical of some details of the existing rules, practices or attitudes, one way or the other, and so contributes to controversy. Your own proposals are all fully within the scope of those debates.[14]

This response reveals one prevailing attitude about the boundaries of legitimate debate in the Netherlands regarding euthanasia. The Dutch have pushed the boundaries of legitimate discussion far and beyond most countries in the world to include psychiatric and demented patients as well as children. People who warn against conferring legitimacy on such problematic issues are often written off. During my interviews I heard time and again dismissive attitudes regarding Koerselman, ten Have, Rutenfrans and Jochemsen, describing them as "religious fundamentalists," "biased," "not rational," and "unable to construct a logical argument." These dismissive remarks led me to believe, quite mistakenly, that Jochemsen is a Catholic, whereas in fact he is, according to his testimony, "a reformed, or Calvinist, bioethicist." I should note that Leenen told me that he wanted to debate with Fenigsen but he never attended any meeting with him. I should add that Leenen had never initiated an invitation to Fenigsen for debate.[15]

John Griffiths was one of the first interviewees to respond. Let me quote at some length the opening paragraphs of his communication:[16]

13 In response, Rutnenfrans wrote: "Den Hartogh probably refers to the comparison I have made between the pleadings to liberalize euthanasia in the Netherlands after the Second World War and similar pleadings to accept euthanasia by social Darwinists like Binding and Hoche in the nineteen-twenties in Germany. Of course the writings of Binding and Hoche influenced the nazi-ideology in this respect. This comparison was published as 'Volg het spoor terug' ('Follow the trace back') in Intermediair, 5 december 1986... Besides, the only 'data' I ever presented were the results of valid researches." Personal communication on August 22, 2001.

14 Personal communication on August 27, 2000.

15 Private conversation with Leenen in Helsinki on August 8, 2000; Leenen's letter dated February 2, 2001.

16 Personal communication on July 10, 2000.

> In general, what you write certainly includes some interesting observations, but is so patently tendentious that I cannot regard it as an interesting contribution to the literature. You claim in the introduction to have "thoroughly" read the "vast literature," whereas in fact you obviously have read only the rather limited literature available in English. And how you imagine that in a month of interviewing you are in a position to pontificate to people who have been working on these problems for years and know a great deal more than you do, is a wonder to me. You leave me with about as much respect for your views as you would have for someone who, knowing no Hebrew, interviewed a number of people in Israel (some of whose English is limited) over for example the Palestinian problem and then wrote an article on the subject in which he rather unashamedly devoted as much attention to a small and highly emotional group of religiously-motivated people as he did to serious supporters (some of them rather critical) of Israeli policy (actually, the degree to which the opponents of Dutch policy are closet Catholics is one subject on which your paper opened my eyes).
>
> This country consists of sensible, levelheaded, decent people, on the whole, who have thought seriously about these matters, and have reached a very high degree of societal consensus. They seem to me to be unlikely to be as far wrong as you imagine. And it seems to me remarkably arrogant on your part to imagine you see far more deeply into the problems than they do.
>
> As you know, I am one of the most severe and consistent critics of Dutch policy, although I do not share the ideological objections of the critics you choose to rely on. But you seem either not to have read, or not to have understood, what we write in chapters 1.4 and 7 [of *Euthanasia and Law in the Netherlands*]. Namely, despite all the criticism one can have of the inadequacy of Dutch control, the situation here is *much better* than it is elsewhere, where people like you think everything is just fine so long as there is a blanket prohibition [I should note that I never supported blanket prohibition. RCA]. Did you read recently of the doctor who killed some 140 women in London? Have you read about the practice of 'terminal sedation' (no control, no request from the patient required) in the US? In short, a criticism of the effectiveness of regulation in the Netherlands, especially when made by an outsider, is necessarily a *comparative* matter. I did not come across a single sentence of comparison in your paper.

In a similar vein, Heleen Dupuis concluded her response by saying:[17]

> My reaction to your book is, in summary, that you are of course entitled to your and your country's own moral views, on which I do not want to comment. But the question remains, why you are so eager to comment on ours? I feel here a kind of moral imperialism, which I really do regret. I can live in a world that rejects euthanasia, and I expect others to be equally tolerant!

Let me say the following in response. When Margaret Battin published her *The Least Worst Death* in 1994 Griffiths, Dupuis and the majority of the Dutch scholars did not attack her for her lack of understanding the Dutch language and culture. Instead, the Dutch euthanasia establishment embraced her views that, on the whole, justify and condone the policy and practice of euthanasia in the Netherlands, and they often quote her writings. On the other hand, Carlos Gomez, who wrote a thoughtful critique of the Dutch policy and practice of euthanasia, was subject to considerable criticism. Griffiths described Gomez's book, *Regulating Death*, based on a one-month research in the Netherlands, as "scientifically irresponsible." To the reader who might dismiss this criticism as exaggerated, Griffiths advised that he would do well to ask himself the following:

> ...how much confidence he would have in the conclusions of - say – a Japanese doctor who studied some controversial medical procedure in the United States by interviewing a handful of American doctors with whom he happened to come in contact about a small number of cases these doctors had been involved in several years earlier... Not speaking any English, our hypothetical Japanese researcher conducted the interviews through an interpreter. Based on the interviews (and without being able to read the American literature on the subject) the Japanese researcher felt able

17 Personal communication on July 25, 2000.

> to make vigorous assertions not only about what American doctors generally do in such cases but also about what influence this has on the patient involved.[18]

Interestingly, when Griffiths (who was born and raised in the United States) related to Keown's penetrating criticisms of the Dutch policy and practice of euthanasia, he said that Keown's argument could be considered painful for the Dutch, but is otherwise a solid one.[19] Griffiths acknowledged that the Dutch simply tend to ignore foreign criticisms, and that this dismissive attitude is "unfortunate."[20]

It is a vexing question to what extent people from one culture are able to evaluate another culture. I have never claimed my expertise or knowledge about the subject matter of this book to surpass that of any of the interviewees. Although I studied the Dutch euthanasia policy and practice for twelve years now I humbly acknowledge that they are far more qualified to comment on the conduct of euthanasia in the Netherlands. I think that their reflections are fascinating and should be of interest to people who are involved in the area, whether they are Dutch or wish to institute euthanasia in their own respective countries.

The line of criticism adopted by Griffiths and Dupuis is not strange to me. A few years ago, when I embarked on the study of questionable cultural practices in Israel, I was attacked on similar grounds. Some of my Palestinian colleagues objected to my studying the practices of female circumcision and murder for family honor in the Bedouin and Palestinian communities,[21] claiming that I do not understand these communities and that I should restrict my studies to Jewish practices. Likewise, when I studied the practice of halizah among the Jewish communities in Israel, some orthodox rabbis did not take well to the idea of a secular person addressing this issue.[22] One prominent scholar in Jewish law, now Justice of the Supreme Court, suggested that I should let the rabbis struggle with the issue.

In short, there are many sensitive toes and many interested parties who do not like "outsiders" treading on them. If we are to follow those "friendly" suggestions, then scholars will restrict themselves to studying only theoretical matters and lofty ideas, while leaving consideration of the more concrete issues solely to those who are heavily involved in them. "Outsiders," however, can bring refreshing modes of thought, offer intelligible critiques and highlight concerns that "insiders" overlook, underestimate, or are reluctant to address. You don't have to be a woman to denounce chauvinism. You don't have to be a Palestinian to condemn the murder of women who "sinned" by merely talking to a man who is not a member of her family. You don't have to be a Hindu to condemn widow burning. You don't have to be African-American to denounce the era of slavery as a terrible chapter in the history of the United States. You don't have to be either German or Jewish to study the Holocaust. You need not be from the United Kingdom to scrutinize the Northern Ireland conflict, or a Chinese citizen to research Mao's "cultural revolution." Many scholars around

18 J. Griffiths *et al*, *Euthanasia and Law in the Netherlands*, *op. cit.*, p. 25.

19 *Ibid.*, at 26.

20 *Ibid.*, pp. 28–29.

21 Cf. R. Cohen-Almagor, "Female Circumcision and Murder for Family Honour among Minorities in Israel," in Kirsten Schulze, Martin Stokes and Colm Campbell (eds.), *Nationalism, Minorities and Diasporas: Identities and Rights in the Middle East* (London: I.B. Tauris, 1996): 171–187.

22 *Halizah* is a Jewish ceremony in which a Jewish widow who has no children releases her brother-in-law from his religious obligation to marry her. Cf. R. Cohen-Almagor, "Israeli Democracy, Religion and the Practice of *Halizah* in Jewish Law," *UCLA Women's Law Journal*, Vol. 11, No. 1 (Fall/Winter 2000): 45–65.

the world examine the American First Amendment, and sociologists and anthropologists from different countries travel to study cultural rituals of other nations. Substantial studies in the fields of history, sociology, anthropology, political science, law and international relations (this is not an exhaustive list) would have been dismissed if people were to adopt Griffiths' reasoning. There were, and there are, many scholars who have enriched the world with their perceptive thinking, despite deficiencies of language. It may even be argued that sometimes remoteness from the subject-matter is a virtue, not an obstacle. This study is not designed to destroy an established movement, but rather to provide constructive criticism. Hopefully, the euthanasia movement in the Netherlands and in the world might benefit and learn from it.

PHASE III: UPDATES

A. PRELIMINARIES

On April 10, 2001 the Dutch Upper House of parliament voted to legalize euthanasia. It was appropriate to address this issue again and ask the interviewees to reflect on the new act and to give their initial appraisal. It was, of course, too early to judge the impact of the act but I wanted to have preliminary observations of the interviewees and to see to what extent they retained, or changed their views since our last communication two years ago.

I also inquired about their assessment of the work of the regional committees. The interviewees brought to my attention some issues that they saw as troublesome. Most of them expressed an opinion regarding two court cases – *Brongersma* and *Van Oijen* – that were on appeal during my second visit to the Netherlands in June 2001. Some mentioned improvements in palliative care and in consultation.

In April 2002 the euthanasia legalization process came to conclusion when the law was approved by the Senate, making the Netherlands the first and – at that time – only country in the world to permit mercy killing. I returned to the Netherlands for the third time, this time conducting ten further interviews. This update summarizes the concerns voiced by the interviewees in June 2001 and in April 2002.

B. ON THE NEW ACT

The first question was: What do you think about the new act? What are the positive aspects of this act, if any? What are the negative aspects of this act, if any?[1]

Henk Jochemsen and Frank Koerselman were the first to respond to my query via electronic mail. As expected, their view of the act was negative. Jochemsen referred me to a recent article in which he voiced several objections to the act. First, the decriminalisation of intentional killing by physicians constitutes, in itself, a serious violation of the legal protection of the life of all citizens. Moreover, whenever the committee rules favorably on a case by deeming an act of killing legal, the Public Prosecutor's ability to monitor physician conduct will be compromised because the Prosecutor will not even see the report of the physician involved in the case. Furthermore, it is likely that cases in which the legal requirements have not been fulfilled will go unreported, as is the case now. Adequate control will continue to be impossible. Chris Rutenfrans expressed a similar objection to the committees, arguing that this institution is too soft and that the public prosecutor should inspect directly the euthanasia reports without any mitigating buffers.[2]

1 Personal communication sent to interviewees on May 15, 2001.

2 Evert van Leeuwen who is a member of one of the regional committees said in his interview (June 20, 2001) that doctors do not have the wrong intent. There is no reason to suspect that doctors kill patients for nothing. Henceforth the committees are lenient and tolerant with regard to physicians' conduct. Gerrit Kimsma said (interview on June 21, 2001) that the committees' underlying assumption is that physicians are acting in good faith.

Second, Jochemsen argues that patients afflicted with terminal illness or unbearable suffering will now have to justify *not* asking for euthanasia. At the same time, legalization will tend to undermine the efforts and creativity of those committed to providing palliative care to terminal patients. Such unintended outcomes seem to Jochemsen inevitable in a healthcare system characterized by increasing costs and rationing.

Third, the new law will lead to a broader acceptance and increased practice of euthanasia, which will change the nature of the physician-patient relationship as well as the character of terminal palliative care. Because the physician's role and the extent of her competence is regulated by law, such a fundamental change in the physician's competence concerns society as a whole and cannot be considered a private matter for patients and physicians.

Fourth, accepting euthanasia of 12 year-old youngsters and older seriously overvalue their ability to make such gruelling decisions.[3] Furthermore, legalizing advance directives is likely to foster a broadening of the requirement of "unbearable suffering" to "loss of dignity" as well as pressure on the physician to terminate a patient's life when a patient has become severely demented, especially when the patient's family insists on doing so. Such a practice might lead to a blurring of the distinction between voluntary and involuntary euthanasia. It is no wonder that the Dutch Association of Nursing Care Physicians had voiced their unhappiness with this part of the proposal. The Association said that its members would not see ADs for demented patients as binding.[4]

Jochemsen concludes that although the responsible ministers have admitted during the debate in parliament that a physician who does not want to perform euthanasia for a patient insisting on having it is not obliged to formally refer the patient to a colleague who may be willing to do so, in practice physicians will feel pressured to either perform euthanasia themselves or refer patients to a colleague. If they refuse both they may run into trouble unless they have indicated in an early stage of the terminal phase of the disease that they object to performing euthanasia. Furthermore, healthcare professionals who reject euthanasia will likely find it difficult to obtain jobs in certain areas of the healthcare system.[5] Chris Rutenfrans made a similar point, arguing that when the law says that physicians may provide euthanasia, the consequences are that physicians who object to euthanasia can base their objection only on conscience; they cannot base their objection on law. Their position is attenuated.

In turn, Koerselman wrote in a similar fashion that the most devastating aspect of the new law is the fact that the purely subjective feeling of "unbearable suffering" and the mostly subjective notion of impossibility to treat or help is enough to accept the killing of people in all phases of their lives in (alleged) illness.[6] In the April 2002 interview, Koerselman specifically related to the issue of advance directives, saying that he objects to abide by advance directives of demented patients. It is possible to provide psychological help to such

[3] Henk Jochemsen, "Update: The Legalization of Euthanasia in The Netherlands," *Ethics & Medicine*, Vol. 17, No. 1 (2001): 7–12.

[4] Ron Berghmans, Arko Oderwald, Evert van Leeuwen, Ruud ter Meulen, Gerrit Kimsma and Sjef Gevers also emphasized this.

[5] *Ibid.*

[6] Koerselman's personal communication on May 16, 2001. Gevers notes that a problematic issue that the law does not resolve concerns patients who approach their physicians early during the course of the disease from which they suffer, asking for euthanasia because they are fed up with life. Many physicians would be reluctant to provide euthanasia under such circumstances because this would mean significantly extending the understanding of "unbearable suffering."

patients. Proper care should be provided to alleviate suffering. Koeselman testifies that his father in law was demented. He was hospitalized in an institution that does not provide euthanasia. Koerselman was struck by the tolerant atmosphere and the high quality of care until the end. At the same time, Koerselman says that when the concern lies with patients in the last stage of Alzheimer, doctors should avoid providing treatment that is aimed at postponing natural causes of death and thus unnecessarily prolongs suffering.

Koerselman provided a further insight by testifying about his own experience. He was consulted about a sixty-five year old patient, in early stage of Alzheimer, who had no supportive family. The patient received a stroke. The medical team provided him with treatment and the patient was able to function independently. However, then the patient entered into a state of delirium and stopped eating and drinking. The medical team then decided not to treat him anymore, and the patient died. The decision not to treat was made because the medical team wanted to spare him the agony of advanced Alzheimer. Koerselman testified that he was, and am upset by that decision. He thinks that the patient should have been treated further, that the medical team should have provided him with more medical care, not to give up on him.

James Kennedy, Gerrit Kimsma and Ruud ter Meulen added another objection to the law. They said that the law gives the impression that patients can now demand to die in a certain way and that physicians are bound to follow the patients' requests. People think that now there is a euthanasia law, therefore they have a *right* to it. This is the prevailing atmosphere, but this view is wrong. Euthanasia is still a criminal act but a physician will not be prosecuted if s/he follows the Guidelines. The regional committee's role is to decide whether the Guidelines were observed and to press charges if they were not. The pubic, however, does not understand this. Kennedy and Kimsma think that there might be more euthanasia requests, conceiving the practice as a normal medical procedure. The growing demand will be voiced not only by patients but also by their family members. Kennedy, however, does not expect a dramatic rise in the number of euthanasia cases. Most patients want to continue living, and most physicians strive to keep patients alive, but there may be more requests. Kimsma thinks that there might be more demand for euthanasia and widening its scope from patients whose suffering is unbearable to patients who feel that their lives are meaningless.

Koerselman and Kennedy do not feel that the new law would make much of a difference. The situation, which in Koerselman's view is most worrisome, will continue to be so. Kennedy thinks it is better to have a law, but he is far from convinced that the 2001 law is an improvement. This law does not add much clarity to the existing situation and does not resolve the complexities mentioned above. Griffiths also feels that the new law will not change much. However, his view of the existing situation is far more positive. He wrote that the law changes nothing significant as far as the substantive requirements are concerned. The only change concerns written advance requests, "but they are probably legal already, and will never be of any great practical importance." As far as enforcement is concerned, the main change is the stronger position of the regional assessment committees (definitive judgment instead of an "advice" to the prosecutors – "but since that 'advice' in practice was always followed, this change, too, is largely symbolic"). Griffiths hopes that the committees and the prosecutors "will be more sensible (dealing with 'violations' in a casuistic way) than the legislator."[7]

7 Personal communication on May 29, 2001.

Arie van der Arend does not think that the new law will make much of a difference either. He argues that with this new Act, end-of-life practices will be unchanged; in fact, common practices will now have a basis in law and certain practices (regarding living wills and rights of young people – 12–16 years of age) have been made more explicit. A 14 year-old can ask for euthanasia and receive it even if his parents object, provided of course that the medical staff agrees to perform euthanasia. Ending the life of patients/clients – whether on request or not – is still included in the Penal Code as a crime.[8]

A similar response came from van der Arend's colleague, Ron Berghmans though the latter is worried about one aspect of the law, ADs, which Jochemsen also pointed out. Like some interviewees,[9] Berghmans thinks that in many respects this law encodes what already was accepted practice: It is only about euthanasia and assisted suicide as specifically defined in the Netherlands, about the criteria for careful practice, and about the review by the regional committees. The most troublesome part in Berghman's view concerns the legal status attributed to advance directives requesting active termination of life. Unlike Griffiths who thinks that advance directives will never be of any great practical importance, Berghmans thinks it is too early to legislate this issue, as it is controversial and much debated, particularly regarding so-called dementia advance directives. However, developments in practice (particularly in nursing home care) must show what significance this part of the law will have and what problems are brought about.[10]

In April 2002 Ron Berghmans told me of a new report of the Health Council (advisory board on healthcare) on demented patients, saying that ADs should be taken into account together with other considerations like the medical prognosis and family's opinion. ADs are not the decisive consideration. Explicit and clear AD refusing treatment should be respected. But it is very difficult to apply an advance directive on euthanasia. This is because the precondition of unbearable suffering is questionable.

In a similar fashion, Bert Keizer argued that ADs for demented patients are completely unworkable in practice. Patients cannot expect doctors to kill them upon becoming demented. A doctor can never act upon such an advance directive because dementia is not an acute condition with a clear start. ADs could help if, in addition to dementia, the patient suffers from a malignant disease. Then the doctor who knows about the AD may decide not to treat the developing disease. But, of course, abstention of treatment is not killing. Keizer never performed euthanasia on demented patients. All the 21–22 euthanasia cases that he was involved in were somatic patients: cancer, stroke, and neurodegenerative diseases like Steele Richardson (slow Parkinson without any medication available to ease the medical condition).

Rob Houtepen was a member of an ethics committee of a nursing home. Nursing home physicians are reluctant to cease life of demented patients even if they have advance directives to that effect. There is also a position paper of nursing home physicians to that

8 Personal communication on May 21, 2001. Interview on April 2, 2002.

9 John Griffiths, Evert van Leeuwen, Arko Oderwald, James Kennedy, A. van Dantzig and Sjef Gevers.

10 Personal communication on May 28, 2001. For further discussion, see Richard Schulz, Aaron B. Mendelsohn, William E. Haley *et al.*, "End-of-Life Care and the Effects of Bereavement on Family Caregivers of Persons with Dementia," *New Eng. J. Med.*, Vol. 349, No. 20 (November 13, 2003).

effect. Dementia does not qualify as unbearable suffering.[11] Houtepen provided the following response, which I quote in full:

> My answer to your question concerning the new euthanasia law is inspired by certain ideas from the field of sociology and philosophy of law. We should not think of a law as a definitive collection of rights and duties. A law is a step in a process. What concerns me is the quality of the process, the communicative and deliberative nature of the euthanasia practice. As far as I'm concerned we are all together on a path without a definitive goal and without clear-cut boundaries. This makes it essential that all information and all points of view are communicated among all participants in the process. No secrets, no dogmas. So my criterion is whether the law facilitates this process. Generally speaking, I think it will. To me it is important that euthanasia will not be a simple right. This means there is always communication and deliberation involved to get the physician(s) to perform the desired act. I also hope that the law puts more pressure on physicians to report all cases. We should continue to give mild sentences in dubious cases (such as *Brongersma*), as we rely on such cases to have a productive societal debate. Thirdly, the law mostly codifies standing practice, so that's all for the better. Finally, I share Ron Berghman's doubts about the role of advance directives. On the other hand, I would not like to rule out all such cases. Therefore, in conjunction with point one and the lucky fact that nursing home physicians are reluctant, I hope that the new law actually stimulates societal debate about such cases.[12]

Evert van Leeuwen, like some of his colleagues, thinks that all that the law does is to legitimize the existing procedure. He does not think that the law would change much in practice. Those doctors who reported in the past will continue to report. The law will not sway those who failed to report. The law has more to do with the possibility for doctors to perform euthanasia than with morality.

Van Leeuwen and Ruud ter Meulen further explained the status of and the rationale for the new law in the broad legal perspective. The law had to be in accordance with other existing laws and therefore it includes unrealistic provisions, such as ADs. Many physicians would not accept advance directives from demented patients. They would regard honoring such a document as breaching their professional convictions. But from a purely legalistic point of view this provision needed to be included in the law because of the *Patient Determination Act* (WGBO, Wet Geneeskunde Behandelovereenkomst), which holds that a patient can give AD on futile treatment. Once euthanasia is allowed under certain provisions, patients should have the ability to issue ADs. The discussion still goes on and van Leeuwen and ter Meulen call to exclude the possibility of euthanasia for demented patients. They are principally opposed to ADs for such patients, arguing that physicians should never be forced to accept them. Ter Meulen points out that many people completed ADs in which they asked for euthanasia but when they become ill only a small number of people still ask for euthanasia. He is worried that euthanasia might become a sort of an escape route for physicians. Instead of investing in more care, physicians would prefer to carry out euthanasia.

Ter Meulen maintained that there are two other weak points in the law. The first concerns the lack of definition of suffering that adds obscurity and opens room for contrasting

11 Cf. "Euthanasia in the Nursing Home: 'We Had a Problem Not to Let the Other Patients Know What Was Happening?,'" in David C. Thomasma, Thomasine Kimbrough-Kushner, Gerrit K. Kimsma, and Chris Ciesielski-Carlucci (eds.), *Asking to Die*, where the author says that nursing home physicians are a little more conservative than the general practitioners "and we tend to protect our patients more because they are weaker. The general practitioner ascribes enormous responsibility to the patient's own health behavior. As nursing home physicians, we are much more protective" (p. 351).

12 Personal communication on May 29, 2001.

interpretations. The second concerns the role assigned to the regional committees. Because they assume a legal role, physicians might not trust them and will not co-operate with them. Indeed, an open question is whether the level of reporting will be improved.

In turn, Gerrit Kimsma is worried about the potential implications of the law that might worsen the existing situation. He would prefer to see the regional committees working for two years before rushing to legislation (in the 1999 interview he spoke of ten years, see *Chapter 11*), arguing that the Netherlands does not need the law. There was a political need. It was the politicians who were pressing for legislation. Kimsma, it seems, would prefer to continue working in the shadowy area that allowed wide latitude for doctor's discretion.

Kimsma maintains that people underestimate the implications of the law that in his opinion its wording is far more liberal than what Minister of Health, Welfare and Sport, Els Borst, intended. In essence, the involvement of politicians increased the polarization of the issue. The provision of ADs is introduced in wording that makes it possible to commit euthanasia of demented people. Presently, there is a grey area. GPs do not end patients' lives if there is no communication between the parties. Now the law, with its recognition of ADs, makes euthanasia possible under these circumstances.[13]

Kimsma foresees more legal cases as a result of this provision. Appeals might be brought to the courts by dissenting family members complaining about the ending of life of their demented love ones. They may claim that the precondition of unbearable suffering was not fulfilled. He explains that a physician may appeal to ADs when no communication with the patient is possible. But this should not be seen as substitute for "unbearable suffering." Furthermore, the work of the regional committees would become more complicated because when there is an AD and it is impossible to communicate with the patient, there is no need to satisfy the consultancy requirement. The consultant cannot affirm the patient's wishes.

However, in response it can be argued that, generally speaking, it is better to legalize than to leave the matter unclear. The euthanasia procedure is stricter and clearer than before. Before the law the Guidelines spoke of patient's hopelessness, but there was no requirement to explore treatment alternatives. Now, according to the law, alternatives for treatment should be explored prior to euthanasia. Second, the Guidelines did not explicitly require physicians to discuss the disease diagnosis. Now physicians are required to provide patients with clear information about the medical situation and prognosis, what symptoms are expected and what the prospects are. Third, the Guidelines now specify that euthanasia should be performed in a medically appropriate way. There are also clear specifications as to what medication should be used in mercy killing. And fourth, the assessment of euthanasia has shifted to the regional committees. Theoretically this is an improvement because the policy distanced the prosecution from the physicians.

Jaap Visser of the Health Ministry testified that the government was happy that the parliament approved the law. He voiced hope that physicians will trust the new law. At the same time Visser acknowledged that it was impossible to answer all the questions. Thus, for instance, the relationship between euthanasia and palliative care will be examined in a new study.

13 Kimsma would like more restrictive ADs. Like all colleagues who expressed an opinion on this issue he argues that it is not a solution to end the life of a happy, demented person. ADs should be concerned with ceasing of treatment, but not with euthanasia. The problem is that for the purpose of euthanasia, the patient being in the terminal phase of the disease was never a condition, and if you cannot assess "unbearable suffering" then you might end up ending the life of a demented person.

I asked him about the provision regarding children, and Visser had no reservations. He explained that the provision is in place and correct. Between the ages 12 and 16, parents have to agree to euthanasia for their children. These incidents rarely occur. In practice, Visser explained, these incidents will always be decided through consensus between the patient in question, his/her parents and the attending physicians.

A. van Dantzig contends that although he believes euthanasia is a medical matter, not a criminal one, he does not object to the law. He thinks it is a progress but he would prefer another system of decision-making. He reiterated what he had said in the 1999 interview, that euthanasia should be conceived as a medical act, like any other medical act, subjected to quality control. To his mind, criminalizing of euthanasia is secularizing of a sin. Then he added that the law does not mean too much. Only 50 percent of cases are reported. In practice, the law did not change much.

H.J.J. Leenen had chosen to reflect on the rationale of and the need for such a law rather than on the content of the law. He assumed that, on the whole, the euthanasia practice was working just fine, hence there was no need to revise the existing practice but only to anchor it in law. He explained that the more euthanasia became generally accepted, the more legislation to change the existing prohibition in the law (Article 293 of the Penal Code) was necessary. Recourse to necessity was only intended for exceptional cases and was no longer appropriate given the number of requests. Leenen maintained: "Moreover, it undermines the authority of the law when the population is no longer behind it and practice is no longer in agreement with the law. Because moral convictions had changed, the existing prohibition of euthanasia had become outdated."[14] I agree with this. I also think that the grey line between what was legal and what was not obscured the issue and did not serve the patients and the somewhat confused doctors. Issues concerning life and death should be discussed in the open light, not in the shadowy area of the law.

Leenen went on to write that the new law bridged the gap between the moral convictions of the population and the old law. "In a democratic society it is not *a priori* dogma but the will of the people" that has to determine what is permitted and what not. He reiterated, and I agree, "Decision-making by the parliament is the only adequate instrument to regulate a social phenomenon like euthanasia." Without adequate legal regulation such protection is lacking "as can be seen in several other countries."[15]

Leenen thinks that the law protects the rights of patients and doctors, but I beg to differ. Leenen writes that the law "establishes a mechanism of review and control"[16] and I would argue that the law establishes an insufficient mechanism of review and control. Leenen argues that "the Dutch law is an example of a clear regulation to steer euthanasia in the right direction"[17] and I would say that it is only one step in the right direction. The law anchors the existing practice and confers it legitimacy, but the euthanasia practice suffers from many flaws that the new law is not likely to amend. The Netherlands requires far more extensive and specified legal regulations to remedy the practice.

To conclude, many interviewees do not think that the new law will make much of a difference. Many of them are quite rightly worried about the legalization of advance

14 Letter dated May 29, 2001.
15 *Ibid.*
16 *Ibid.*
17 *Ibid.*

directives (ADs), especially when demented patients are concerned. I wrote elsewhere[18] that ADs are often made without the opportunities for full informed consent. On what basis did the patient make the decision that she would not want to be treated? The decision not to receive treatment should be based on a clear understanding of the situation. It is essential that the patient really understands the disorder, the available alternatives, and her chances and risks. This can be rather complicated when the physicians themselves do not have a clear picture about the condition and cannot provide a reliable prognosis.

Furthermore, there are valid concerns about the patient's opportunity to change her mind, the potential for scientific developments, and the clarity of the advance directive. In a patient who is mentally alert and who makes a decision not to receive treatment, there is always the opportunity for the clinician to consider with the patient the reason for the decision in light of the particular clinical features. The clinician also has the opportunity to discuss with the patient why she does not want treatment. Is it fear of pain, loss of dignity, concern for others, etc.? In these circumstances, the clinician is in a better position to ensure that the patient has thought out the decision carefully. This possibility is removed when faced with an AD and an incompetent patient.[19]

As for the potential for scientific progress, competent patients have the advantage of making their decision based on up-to-date knowledge. Advance directives may be made many years prior to the time of their implementation, a period during which new treatment or changes in quality of life opportunities may have occurred. In addition, ADs are not always clear about the patient's intentions. Keith Andrews testifies that he saw an AD stating that if the person developed severe brain damage she would not want to continue living. There was no statement as to whether this decision was to be made on the first day or after a period of several days/weeks/months to give the patient opportunity of recovery. The general statement about severe brain damage gives wide latitude for widely differing views, even among clinicians experienced in the management of brain damage.[20]

The procedure of advance directives reflects competent values and interests in circumstances where they might no longer be applicable. If the patient is incompetent at the time the procedure becomes effective, withdrawal or withholding of medical treatment constitutes a form of imputed consent. The situation of the incompetent patient is viewed through the lens of her prior competent self rather than her current state. It is usually assumed that the justification for giving the competent person power over decisions when she is incompetent is that the competent person is best situated to identify what those future interests will be. The problem, however, is that the patient's interests when incompetent are no longer informed by the interests and values she had when competent. The values and interests of the competent person are no longer relevant to someone who has lost the rational structure on which those values and interests rested. Although still the same person, the patient's interests change radically when she becomes incompetent. Thus there is a possible conflict between past competent and current incompetent interests.[21]

[18] R. Cohen-Almagor, *The Right to Die with Dignity: An Argument in Ethics, Medicine, and Law* (Piscataway, NJ: Rutgers University Press, 2001), chap. 5.

[19] Keith Andrews, "Euthanasia in Chronic Severe Disablement," *British Medical Bulletin*, Vol. 52, No. 2 (1996), p. 287.

[20] *Ibid.*, p. 287.

[21] John A. Robertson, "Second Thoughts on Living Wills," *Hastings Center Report*, Vol. 21, No. 6 (November-December 1991): 6–9, at 7.

To resolve the conflict between past competent and current incompetent interests, it is suggested that instead of simply enforcing all prior directives, doctors, families, and other people involved in the care of incompetent patients should be able to examine whether the patient's interests would best be served by actions contrary to the AD, in situations in which the incompetent patient appears to have an interest in further treatment and life.[22]

Summing up the considerations that are important in decisions on death with dignity, I would say the following: if the patient makes advance directives in the form of a living will, DNR order, a letter, etc., that she would wish to continue living, no matter what, and we have no reason to believe that the patient changed her mind, then we should respect that wish.

In the event the patient prepared an AD saying that she would prefer that all treatment be terminated upon reaching the last stage of their incurable disease, and we are uncertain about the patient's present wishes because, for instance, she is incompetent and the attending physicians think that the situation is irreversible, then we should respect the AD and let the patient die. For persons who prepared ADs asking to die upon reaching a certain situation, death is not the worst situation one can be in when compared to being on the verge of death and then stabilized without hope of ever really getting better. Patients who suffer from incurable diseases (like cancer) may feel that their lives become transient and that the thought of death brings them more comfort than alarm. They may feel that their dignity, their autonomy, their humanity are better served by letting them die. The patient's wish must be respected. This is especially true if the patient has emphasized beforehand that her dignity cannot be separated from consideration of her loved people. For some patients, knowledge of the anguish their condition imposes on their families is such a heavy burden that they prefer to die and not be remembered in their diminished condition. This, of course, is not the sole consideration but a significant additional consideration that needs to be taken into account.

If no ADs are available, we should ask the advice of the patient's beloved people who should know the patient better than anyone else. If the patient's loved ones believe the patient would want to be kept alive, then we should respect their decision, even if the attending physicians disagree. In the event that the patient's beloved people wish to withhold treatment and the attending physicians think there is still a hope of recovery, then we have to respect the physicians' decision. The patient's best interests require erring on the side of life.

In the event that the patient's beloved people and the attending physicians believe the patient's condition will only deteriorate, and that that condition negates her dignity, the best interests of the patient require allowing the patient to die. I repeat: the best interests of the *patient,* not those of the family or other beloved people, the physician, or the hospital or the society at large.

C. ON THE WORK OF THE REGIONAL COMMITTEES

Next I asked about the regional committees. In 1999 the committees just started to work. I inquired what the interviewees think about their work now that they had more time to reflect on their conduct. This is of special significance due to the important role they play in the policy and practice of euthanasia nowadays.

[22] *Ibid.,* p. 8.

According to the committees' official report, during the year 2000 the five regional committees reviewed 2,123 cases. Of these, 1,801 cases were performed by GPs, 278 by medical specialists in hospitals and 44 by physicians in nursing homes. There were 1,893 cancer patients, 10 had AIDS, 28 suffered from heart diseases and damaged blood vessels, 51 had mental illnesses, 45 had lung diseases, and 96 suffered from other illnesses.[23]

Several interviewees argue that committees are doing good, serious and important work, expressing satisfaction with their conduct. Their feeling is that the assessment process has been improved. At the same time they noted that the expectation of a higher level of reporting was not fulfilled. They hope that the new law will help increase the level of reporting.[24] There is scope to think that physicians do not notify the committees because they probably do not fulfill all the requirements. Possibly they do not fill the paperwork adequately. Some physicians think that euthanasia is a private matter between them and their patients, and that no outsiders should inspect their work. If at all, so they think, inspection should be done by the medical profession, but certainly not by lawyers. Furthermore, the regional committees are seen by many physicians as bodies that were constructed by the government. One might suspect that as soon as the government will become unpopular, physicians might be less willing to cooperate with the committees because they identify them with the government.

Evert van Leeuwen provides an interesting hypothesis contesting the data that the level of reporting did not improve. He argues that the level of reporting is actually higher than fifty percent. His guess is that eighty percent, maybe even a higher percentage of euthanasia cases, are reported by GPs. This is because no information on unreported cases is revealed by other sources, like patients' relatives or pharmacists. This suggests, argues van Leeuwen, either a plot between all the concerned parties to hide cases of euthanasia, or problems in the research methodology. To further examine this hypothesis, van Leeuwen suggests including more questions in the coming survey (for instance, how many times do you administer muscle relaxation?) to double-check the physicians' conduct and to inspect their behavior and decision making process.

On the other hand, James Kennedy thinks that the committees have been a disappointment, a failure. One of the main aims, arguably the chief reason, in establishing them was to increase the number of reported cases of euthanasia. In practice, the number of reports has decreased slightly but significantly.[25] One interpretation of this statistic is that the number of requests went down. Another interpretation is that the committees did not calm physicians' fears of prosecution but quite the opposite. Physicians might be more afraid of the regional committees than of the public prosecution because they are specialists. It is difficult to trick or mislead them. They are trained in the area of mercy killings, know the subject matter arguably better than the physicians, hence physicians do not want to face their tough interrogation and do not report. Kennedy maintains that there are good people on the regional committees. They are not there to hang doctors. They take their business seriously

[23] Ministerie van Volksgezondheid, Welzijn and Sport, *Regionale toetsingscommissies euthanasie,* Den Haag, Jaarverslag 2000 (in Dutch).

[24] This was the view of Jaap Visser, Sjef Gevers, Arko Oderwald as well as of committees' members Gerrit Kimsma and Evert van Leeuwen. Bert Keizer said that he had a very limited involvement with their work. He reported one case to the regional committee and received a note saying that his conduct was fine.

[25] Chris Rutenfrans wrote that in 1999 there were 2216 reports of euthanasia whereas in 2000 the number of reports was 2123. Personal communication on August 22, 2001.

and approach the issue in the most humane and professional way; but the question remains of the extent to which euthanasia can be regulated given the low level of reporting.

Gerrit Kimsma would like to expand the mandate of the committees to include pedagogy. He argues that the regional committees are currently concentrating on examination of procedures. Kimsma thinks that the pedagogic function on the process is sorely missed. Potentially, the committees could have provided excellent feedback to physicians about their conduct. Unlike the courts that are asked to assess the legality of the conduct, the committees also could have provided feedback to physicians, helping them to improve the decision-making process. Unfortunately, argues Kimsma, this is not the case.

Kimsma sits in his regional committee seven months a year and reviews some 50 cases of euthanasia each month. He recalls three incidents in which the committee recommended prosecution. In an additional 2 or 3 cases, the committee advised opening an investigation because it did not have a clear picture of the physicians' conduct. The prosecution decided to drop the charges in two cases. The third case was under consideration. Kimsma does not agree with the prosecution on those two cases, thinking that a legal procedure is needed to clarify the limits of the euthanasia practice. He thinks that there is still a long way to go before the Netherlands should be satisfied with the practice of euthanasia.

I asked about the number of cases that were brought to the courts upon the recommendation of the committees. Jaap Visser argued that until April 2002 there have not been any court cases. During 2001-2002 there were three incidents in which the committees voiced concerns regarding the conduct of doctors, but the public prosecutor decided not to press charges in any of those occasions.

Visser's impression was that physicians do not like the regional committees and are reluctant to report to them. This is because many of them feel that the people on the committee are more judgmental than the public prosecutor. Visser thinks that doctors are more aware of the importance of a careful decision making process. They know the criteria for careful conduct, and what is required of them to do. At the same time, Visser testifies that the government holds a very positive view of the work of the regional committees. Also, the Dutch Medical Association has a very positive view of their work.

There is a delicate tension in the work of the committees. If the committees return to doctors performing euthanasia and ask questions, then they might be perceived as too tough on doctors, and then doctors will be reluctant to report. Now the committees are at the stage when they try to build trust between them and the doctors. So the general climate is to encourage physicians' cooperation. The committees may ask questions but they try not to step on doctors' toes. Due to the work of the committees, doctors have more information about the Guidelines: how they should perform euthanasia, encourage careful practice, making doctors think about their own reasoning.

As said before, the main function of the committee is to increase the level of reporting. I raised the issue of reporting also during my third round of interviews in April 2002. Three interviewees estimated that the level of reporting is around fifty per cent.[26] Jaap Visser said that in 2000 the number of euthanasia cases reported to the regional committees was 2200. In 2001 there were less than 2100 reported euthanasia cases. Visser did not have a sufficient explanation for this reduction of cases, whether there were less euthanasia cases, or fewer

[26] Arie van der Arend, Henk Jochemsen, and Ron Berghmans.

cases were reported. He estimated that the number of reported cases is now around sixty percent.[27]

D. FURTHER CONCERNS

Some of the interviewees with whom I spoke during my second visit to the Netherlands initiated discussion of the two court cases that were on appeal at that time. Those were the *Brongersma* and the *Van Oijen* cases.

In 2000, a court in Haarlem considered the case of Edward Brongersma, an 86-year-old man who had first made a written euthanasia declaration in 1984. He had expressed his wish to die to his GP, Dr. Philip Sutorius, on at least eight occasions. A month before his suicide in April 1998 he said that death had "forgotten" him, his friends and relatives were dead, and he experienced "a pointless and empty existence."[28] A second medical opinion confirmed that he was suffering hopelessly, and a psychiatrist said he had no treatable psychiatric illness. However, the public prosecution had called for the GP to be given a three month suspended prison sentence. The prosecution recognized that Dr. Sutorius fulfilled all the legal criteria but one: "hopeless and unbearable suffering." Therefore, the patient's request should have been refused. The court did not penalize Dr. Sutorius, saying that the patient was obsessed with his "physical decline" and "hopeless existence" and therefore was suffering "hopelessly and unbearably". A spokesman for the Royal Dutch Medical Association reacted to the court judgment by saying that the definition of "unbearable suffering" had been stretched too far and that "what is new is that it goes beyond physical or psychiatric illness to include social decline." Justice Minister Benk Korthals had said that being "tired of life" is not sufficient reason for euthanasia.[29]

That was also the prevailing view among those interviewees who raised this court case for discussion.[30] They rightly contended that it is medically wrong to prescribe euthanasia on the grounds that the patient is fed up with life. Chris Rutenfrans noted that Brongersma was not ill, did not suffer but still his GP performed euthanasia on him. The Guidelines speak of unbearable suffering and here it was not fulfilled. At least, Brongersma did not suffer physically. Rutenfrans rightly argues that *Chabot* was the precedent that enables this development. Now everyone can claim mental suffering and ask for the doctor's assistance.

In a similar fashion, Ruud ter Meulen said that many doctors were shocked by this case. Doctors think that people who are tired of life should not be their concern. Physicians

27 In 2001 the five regional euthanasia assessment committees received 2054 reports, compared with 2216 in 1999. The Ministry of Health says the reasons for the current decline in reported cases are unclear, and it awaits the results of the next study, by Gerrit van der Wal and Paul van der Maas, in 2003. The chairwoman of the assessment committees, Reina de Valk, argued that the decline in reporting could be due to doctors' increasing knowledge of palliative care, promoted by the Ministry of Health since 1997. She also argued that a recently completed national advisory network of 465 doctors belonging to Support and Consultation over Euthanasia in the Netherlands (SCEN) could have led to a decline in euthanasia. Cf. Tony Sheldon, "Reported Euthanasia Cases in Holland Fall For Second Year," *BMJ*, Vol. 324 (June 8, 2002): 1354.

28 Tony Sheldon, "Dutch GP Cleared after Helping to End Man's 'Hopeless Existence'," *British Medical Journal*, Vol. 321 (November 11, 2000), 1174.

29 *Ibid.*

30 Evert van Leeuwen, Gerrit Kimsma, Rob Houtepen, Ruud ter Meulen, Chris Rutenfrans, Henk Jochemsen and Arko Oderwald.

protested against Sutorius's conduct. The questions revolve around the issue of suffering: What is suffering? Who determines suffering? What are the criteria for suffering?

Frank Koerselman thought Brongresma was a clear case of a treatable psychiatric condition. Brongresma was depressed. His neighbors terrorized him because of his past as a pedophile. People threw stones at his windows. People shouted at him and insulted him. All this made Brongresma very depressed. A psychiatrist was consulted and said that Brongresma was not depressed and that he made the euthanasia request with a clear mind. However, Koerselman begs to differ. He believes that Brongresma could have been treated and helped. Were Koerselman the consultant, his advice would have been to provide him care and treatment, and try to solve the problems in his neighborhood.

Arko Oderwald thinks that Sutorius's conduct in the *Brongersma* case was absolutely wrong. Brongersma talked to Sutorius for hours on this but Sutorius should have refused. Brongersma had no terminal disease. Sutorius should have said to his patient that he has his full sympathy but he should not request his physician to kill him. No physician has the right to conduct euthanasia simply because the patient asks for it. Oderwald is not convinced by the argument that Brongersma was mentally suffering. After all, Brongersma himself did not try to commit suicide.

Gerrit Kimsma said that the patient wanted to die but he did not suffer unbearably. Nevertheless nobody would wish to convict and penalise Sutorius. Kimsma explained, "we have a very liberal abortion law allowing abortion to any woman for any reason. We now progress with euthanasia in the same direction, conceiving this as a civil rights issue." This is the result of autonomy that rules supreme.

I asked Paul van der Maas about this and the *Van Oijen* case. Van der Maas wrote: "To be frank, I have not yet formed my opinion about these specific cases. I think the issue of assisted suicide in the absence of severe suffering is however very important, and deserves public debate. We will include the issue in our next nation-wide study."[31]

In turn, Gevers, Berghmans and den Hartogh have mixed views about the *Brongersma* case. Gevers explains that the Appeal Court asks further expert opinions before deciding the issue. True, the patient had no somatic illness but he did have a number of somatic complaints associated with his age. His suffering was psychological. He lost all his friends, found no meaning in life. Some argued that he was therefore suffering unbearably. Gevers thinks that the Haarlem Court's decision contained some good reasoning. However, he also shares the public policy concerns. Gevers thinks that the medical profession should address this issue more rigidly. Indeed, the Dutch Medical Association had decided to set up a committee to investigate this issue, and Gevers is curious what the committee will say.

Ron Berghmans is unsure about the case. On the one hand, it is difficult to argue that being tired of life constitutes unbearable suffering, and it is questionable whether the doctor has a role to act in such cases. Berghmans agrees with the two expert opinions that the doctor does not have a place here because his assistance is not intended to alleviate medical suffering. On the other hand, Berghmans could imagine some cases in which old, lonely patients might claim that they are tired of life and their condition might be considered as unbearable suffering. But there need be more than expression of "tired of life". For instance, inability to cope with daily activities, dependency on others, no friends.

31 Personal communication on August 20, 2001.

Den Hartogh elucidates that there are two relevant questions in the *Brongersma* case: (1) Was it a case of unbearable and hopeless suffering? Not all the relevant details are public knowledge, so he cannot make a definitive judgment. But the answer might be yes. (2) Is it necessary for the suffering to be caused by either somatic or mental illness? Den Hartogh thinks it is difficult to delineate those cases clearly. In the *Boomsma* case, Chabot claimed that she was not mentally ill, still in his view a psychiatrist can be properly involved with such a person because he can offer her assistance in overcoming the effects of severe traumatic experience. The disciplinary court did not share his view but considered her a "psychiatric patient". The position of the High Court on this is not fully clear. In the *Brongersma* case, it is obvious that the effects of old age essentially contributed to his suffering, and it is also normal that a person of that age and condition is under regular medical supervision and to this extent is a "patient." But it is also clear that the relationships between Brongersma and Sutorius were largely defined by Brongersma's death wish, rather than by his condition. Should assisted suicide be an option for people who are not patients or are only marginally so? Den Hartogh testifies that he had not decided that question. On the one hand, if the basic reason to grant someone's request for assistance is the fact that the person is suffering unbearably and hopelessly, then the cause of suffering seems irrelevant. On the other hand, "it is not uncommon in law (if not in ethics) to insist on criteria which are irrelevant in themselves, in order to protect the relevant criteria from being undermined. The notion of unbearable and hopeless suffering is inherently a rather vague one." It would be very difficult, almost impossible, for an assessment committee to judge whether the criteria are satisfied, if the symptoms cannot be interpreted in the context of a medical condition. Den Hartogh argues that if the court decides to condemn Sutorius, he would feel sorry for people like Brongersma "who maybe suffering more severely than many cancer patients whose request for euthanasia is granted" without asking too many questions. But, den Hartogh maintains, he would also be relieved that "no more ground is lost to the people who want to grant each person his request as long as it is competently made. The second feeling is at least as strong as the first one."[32]

The most supportive views of Sutorius' conduct were expressed by Jaap Visser and Bert Keizer who felt that Sutorius did his job with conscience. Visser talked with Brongersma many times and realized that there was no more help for him. But this is not a case for euthanasia under the law. "We don't have an answer for such patients whose condition does not fall under the euthanasia Guidelines."[33] In turn, Keizer finds the physician's behavior justified, saying that he himself probably would have gone along with the patient's request. Keizer sees age as a very important factor in euthanasia requests. Brongersma was 86 year old. You cannot say to him: You will grow out of this. Keizer says that he would have stated on paper that Brongersma's physical situation was such that he could not live independently any longer, that he was in complete social isolation. Keizer said he would probably have described the situation as hopeless and by this turn the case into an acceptable case of euthanasia. Old people who are completely dependent on other people, who conceive their own life as hell are entitled to ask for euthanasia, and the doctor is entitled to provide assistance.

32 Personal communication on August 6, 2001.

33 Interview on April 3, 2002. For further discussion condoning Sutorius, see Peter Singer, "Voluntary Euthanasia: A Utilitarian Perspective," *Bioethics*, Vol. 17, Nos. 5-6 (2003):535-536.

Keizer went on to relate the last case of euthanasia (PAS actually, but he doesn't feel the distinction to be very important) he was involved in. An old lady who led a very active life and then suffered a stroke. As a result, she became half paralyzed. She had a hemianopsia (half-sided loss of visual field) that prevented her from reading and watching television. Then, after 6 months striving for rehabilitation, she also broke her hip and subsequently asked to die. After the stroke she did not lose hope and fought but after the broken hip she gave up on life. Keizer testifies that she could have lived like this for years but she asked for euthanasia and he agreed. Keizer agreed because he thought her situation was hopeless. Dependency was too much for her. She was not in pain but emotionally she was suffering. She could not get up, or walk, or get into bed, or visit the toilet without help, or eat alone. All that she could do was to speak, sit and lie down.

Back to *Brongersma,* in December 2001 the Amsterdam Appeal Court accepted the prosecution's petition and reversed the lower court's ruling. Although the court found Philip Sutorius guilty, it still refrained from giving him a jail sentence. Sutorius was found guilty because he did not act for medical reasons, but rather because the patient was tired of life. The court did not sentence Sutorius because he acted out of compassion for his patient and because the court viewed this as a test case by the prosecutor.[34] The lack of sentencing conveys a dubious message to physicians who might find themselves in a similar position.

A year later, in December 2002, the Dutch Supreme Court upheld the conviction of Sutorius. The court said it was clear Dutch euthanasia rules were never intended to cover such cases. The ruling by the nation's highest court insisted that the medical Guidelines for mercy killing should be strictly observed.[35]

The second case, *Van Oijen*, involved an 84-year-old patient in a nursing home who unquestionably experienced unbearable suffering, but there was no request for euthanasia on the part of the patient. The patient had had heart problems and osteoporosis for a long time, and during the last year was increasingly bedridden. Her physician was Wilfred van Oijen who – you may recall – was the physician who appeared on the film, *Death on Request* (see *chapter 9*). Chris Rutenfrans elaborated on this case, saying that according to the reports van Oijen came to visit his patient and saw her unconscious and lying in her dirt. The patient's daughters stood next to their mother and said to van Oijen, who had been the patient's GP for 17 years: "Even a dog you shouldn't let die in this way." Then van Oijen gave her a lethal injection and she died. The patient never asked for euthanasia, and van Oijen failed to consult a colleague and also reported the death of his patient as a natural death.[36] However, his conduct was brought to the attention of the legal authorities and the case came before the court. The Amsterdam Court said that Dr. van Oijen had made "an error of judgment" but had acted "honorably and according to his conscience," showing compassion in what he considered the interests of his patient. Thus, although van Oijen was convicted of murder he was given only a suspended sentence of 5000 guilders for breaching the Guidelines. The

34 "Dutch Doctor Is Convicted in Suicide Case," *New York Times* (December 9, 2001).

35 "Dutch Court Upholds Euthanasia Conviction," *The Associated Press* (December 24, 2002). For further deliberation, see B. Rosenfeld, "Assisted Suicide, Depression, and the Right to Die," *Psychology Public Policy and Law*, Vol. 6 (June 2000): 467–488.

36 Govert den Hartogh commented: "My knowledge of the details of the *Van Oijen* case is insufficient to comment on it. It is clear that he failed to report and such failures are less and less excusable, because one can no longer say that the criteria are unclear. (Maybe van Oijen was fully aware that he did not satisfy them.)" Personal communication on August 6, 2001.

Royal Dutch Medical Association (KNMG) has defended van Oijen's action as having "complete integrity," claiming a "huge emotional gulf" between it and the offence of murder.[37] Legally, what van Oijen did was murder, but it is argued that he conducted the murder with good intentions.

The interviewees expressed mixed views on the *Van Oijen* case. While Ron Berghmans had no clear opinion on the case, others condoned van Oijen's action while still others condemned him. The foremost support was voiced by Rob Houtepen who said that the patient's condition was terrible. She could not control her bodily functions, suffered from severe wounds; her suffering was unbearable and insolvable. Her prognosis was that she would live for one or two days before she dies. In these circumstances, the humane thing to do was to terminate life, even regardless of having a euthanasia request. Euthanasia should always be a measure of last resort, and here it was a measure of last resort. Houtepen maintained that van Oijen's mistakes were insufficiently explaining to the patient the consequences of her policy treatment. She refused to take morphine. Consequently, full palliative care treatment was not provided. Van Oijen should be blamed for not providing the patient adequate information about her disease and treatment choices, that her condition might become unbearable without adequate palliative care. He failed also to be clear with the patient's family about his plan of conduct, his intentions regarding treatment. His pattern of previous judgments shows that van Oijen was very liberal with euthanasia.

Jaap Visser thinks van Oijen took the right decision. The patient was suffering. The nurses did a bad job, left the patient to herself. Therefore, he should be acquitted. His fault was that the patient did not ask for euthanasia, but medically he had done the right thing, what was required of him as a medical doctor. Jochemsen thinks this is a very difficult case. The patient suffered from open wounds and was lying in her dirt. Van Oijen had the feeling that he must act in order to free the patient from her dehumanizing situation. Jochemsen condemns the action as murder but maintains that if society accepts euthanasia, then he could sympathize with van Oijen's conduct. When euthanasia is allowed, providing death can be expected.

On the other hand, Arko Oderwald, Frank Koerselman and Bert Keizer opined that van Oijen practised bad medicine. The patient had no advance directive or living will requesting euthanasia. Still, van Oijen assumed the responsibility. Oderwald thinks van Oijen's license should be revoked. Oderwald testified he does not trust doctors who think they know best.

Koerselman also thought that van Oijen had done a very bad job as a doctor. He violated all the Guidelines for careful conduct. The patient did not ask for euthanasia. Van Oijen did not consult anyone. In addition, he did not follow the Guidelines regarding the drugs to be administered. He injected the patient a muscle relaxant drug, taken from a bottle he used earlier to euthanize another patient. Koerselman was very unhappy that van Oijen was glorified as a hero by the media, portraying him as a humane doctor who did the right thing to relieve suffering. Van Oijen even received a special praise by the Editor of *Medisch Contact*, the official journal of the Royal Medical Association.[38]

[37] Cf. Tony Sheldon, "Dutch GP Found Guilty of Murder Faces no Penalty," *BMJ*, Vol. 322 (March 3, 2001), p. 509.

[38] In an early communication, Frank Koerselman commented on the two cases, *Brongersma* and *Van Oijen*, by saying: "Both of them are, in my view, quite bizarre. Of course I only know these cases from the media. But as far as these [commentaries] are reliable, both seem to be examples of very serious medical malpractice. On the other hand, the official journal of the Royal Dutch Medical Association is treating both doctors as if they were

Bert Keizer found van Oijen's conduct to be completely wrong. Van Oijen succumbed to family pressures. Keizer said that there are various ways to cope with such pressures. The answer to such pressure is not killing of the patient. The patient was unconscious, therefore she was not suffering. Why, then, kill her? Keizer knows van Oijen personally, thinks that he is a good man, but here he made a mistake.

The two cases show that it is possible to compromise either of the absolutely important preconditions of patient's voluntary decision or of unbearable suffering and still the physicians would not be punished. So long as the Dutch think that there was a reasonably good justification to infringe on a certain Guideline, they show sympathy for the physician's conduct even when it involves the killing of a patient. My view of those two cases is highly critical. Instead of investing efforts to improve the patient's living conditions physicians opt to perform euthanasia. In the first case, Sutorius should have refused to discuss the issue of euthanasia with his patient, by this showing Brongersma that this is not an issue to ponder. Undoubtedly the case is the result of a very permissive culture of euthanasia. I never heard of a similar case in Israel, England, the United States or in any other country for that matter. Physicians should not fulfill *all* patients' requests. Sometimes they must withstand their patients' pressure and explain in an unequivocal manner that they must remain loyal, first and foremost, to their conscience. I myself think that physician-assisted suicide is justified under certain conditions, which should be satisfied. These conditions are explained in the *Conclusions*. The conditions were not satisfied in the two cases discussed here.

The second case is also highly disturbing. Instead of investing more time and effort to improve his patient's quality of life, van Oijen simply killed her. Dying patients deserve more care, not a syringe full of poison. No healthcare system should allow physicians to walk into medical departments with a ready-to-inject syringe and finish off patients, even if they are in great suffering. Just imagine that each of the hundreds medical personnel will kill only one patient deemed to be suffering and/or in inhumane conditions. If my opinion were to be heard, I would recommend van Oijen and the Dutch medical profession to declare a mutual divorce.

Chris Rutenfrans further reflected that after the parliament had passed the euthanasia law, on April 14, 2001 Minister of Health Borst declared that she would welcome discussion on introducing a suicide pill for elderly people. Some ten years ago, a judge by the name Drion wrote an article in favor of a suicide pill. Then the suggestion evoked intense discussions and now Minister Borst reintroduced it on public agenda. The idea is to provide a pill for healthy people who are tired of life. By introducing this subject Minister Borst showed her support for Sutorius.

Minister Borst was severely criticized for her suggestion, not only by political parties that are opposed to the government but also by the liberals and social-democrats who form part of the government. In the face of the heated debate, Borst withdrew her proposition. She understood that politically it was not the right moment to pursue the issue. Arko Oderwald explained that the reaction to Minister Borst's proposal on the suicide pill was negative because the Dutch are studying the recent developments in the euthanasia sphere and do not wish to rush further.

holy victims of the Dutch legal system, just because they are prosecuted (until now they have not been sentenced, as far as I know)." Personal communication on July 27, 2001.

Evert van Leeuwen and James Kennedy noted two positive developments that have taken place since 1999. The first concerns palliative care. The government has put a lot of money into developing this field of expertise. Courses in this area are provided to physicians and more research is performed. A lot was done to improve palliative care in hospitals and in medical care at large.

On the other hand, Ruud ter Meulen argues that palliative care is indeed improving in the Netherlands but it is still not integrated well into the nursing care system. It is still underdeveloped. Ter Meulen suggests (and I fully agree) that the Guidelines include a provision for palliative care before considering euthanasia. This would prevent administering euthanasia too quickly. James Kennedy remarks in this context that there is some indication that some physicians perform euthanasia simply because they do not know what else to do. With the growing attention to palliative care, physicians may recognize new vistas that are open to them and not rush to perform mercy killing.

The Ministry of Health, Welfare and Sport announced that the government has set aside NLG 7 million a year for a period of five years to improve palliative care. Some of this money is used for research and some of it to find ways of increasing the scope for its use in the closing stages of life. Six teaching hospitals have set up centers for the development of palliative care. Some of these centers have expert teams of consultants who can answer questions from GPs, nurses and home care providers about patient care.[39]

Henk Jochemsen argued that in the cities, where the attitude to euthanasia is permissive you see a kind of retreat on the part of some physicians because now they have better possibilities to alleviate suffering. In the countryside, where the euthanasia practice did not develop to the same extent as in the cities, palliative care is less developed and the practice of euthanasia seems to expand gradually and the palliative care infrastructure is still insufficient. There is less awareness regarding the importance of palliative care.

In this context I express hope that now that the euthanasia law has passed, the discussion about euthanasia will calm down a bit and leave room for other concerns, no less important. There is a need to improve the care provided to dying patients and to conduct more research on the withholding and withdrawing of treatment. After all, only a small percentage of patients ask for euthanasia.

The second positive development concerns the systematic consultation that is now in place (for discussion on the SCEN project, see *chapter 9*). Leenen and Kennedy argue that the SCEN project is working quite efficiently in most places. In the past, physicians used to consult colleagues when the issue was already decided, a *fait accompli.* The role of the consultant was, accordingly, very limited. SCEN tries to make consultations at earlier points in time and to explore other options, such as palliative care. SCEN consultants are increasingly becoming more qualified in palliative care. They explore other treatment options and verify that euthanasia is, indeed, the only option left. Kennedy maintains that more and more physicians regard consultation with SCEN doctors in a positive light and consult them.

Bert Keizer also has a very positive view of the SCEN project. He thinks SCEN doctors have a broader range of experience, can provide broad understanding of medical conditions, and can point to more comprehensive reasons whether to approve, or disapprove of a request for euthanasia. Keizer testifies from his own experience that he consulted SCEN doctors

39 http://www.minvws.nl

twice, and on both occasions the consultants helped him strengthen his own position regarding the patients' requests. On both occasions, the patients asked for euthanasia, Keizer did not agree and the consultants supported his position.

Ruud ter Meulen serves as a substitute member of the regional committee in his region. He testifies that if the consultant is a SCEN member then he is confident that the judgment is independent, that the consultant accurately evaluated the patient's medical condition. SCEN doctors are also more experienced, more knowledgeable than GPs who are not members of the SCEN program.

Jaap Visser said that the plan is to have a comprehensive network of SCEN consultants all over the 23 regions of the Netherlands. In April 2002 SCEN was available in 16 of them. Physicians are not obliged to consult with SCEN doctors but many approach them. The regional committees said in their annual report that the quality of SCEN consultation is higher than the consultation with a non-SCEN physician. SCEN doctors have special qualifications. They undergo three day course on issues concerning end of life: palliative care, juridical procedure, pharmacology, role playing of patients, their families and the medical staff. They learn how to deal with family pressures. SCEN physicians tend to emphasize palliative care before recommending euthanasia.

Visser further explained that any GP can be on the SCEN list of consultants if he/she asks so. Then they will have to do the course. Some physicians said after the course they did not want to be on the SCEN list because they rather not be involved in euthanasia decisions. SCEN doctors may perform euthanasia and not only serve as consultants. Visser elucidated that the consultant sees the patient and the GP in charge. The process from the first phone call of the GP to the consultant, through the visitation of the patient, talking to the GP, till the writing of the report takes 4 to 5 hours. The consultant spends between 30 to 60 minutes with the patient. Sometimes the consultant asks to see the patient for the second time.

Henk Jochemsen said that some SCEN doctors stated they are now more reluctant to administer euthanasia or recommend euthanasia because they discovered that patients and families sometimes request euthanasia in early stage of the disease, and also that palliative care may alleviate suffering, which is often the motivation for initiating euthanasia requests. In turn, Frank Koerselman spoke of an interview with some SCEN doctors who said that they were disappointed by the communications of patients and families. Patients and families spoke of their right to euthanasia. Patients focus so much on euthanasia that they are not willing to discuss treatment options. SCEN doctors also expressed remorse about their past conduct, now realizing after acquiring knowledge about the benefits of palliative care, that they had euthanized patients prematurely. Some of the patients whom they killed could have been helped with the administration of palliative care. Koerselman maintained that the SCEN program is working well and is expanding throughout the country. More and more GPs are approaching SCEN doctors for consultation.

PART C

CONCLUSION

A. PRELIMINARIES

I came to the Netherlands with mixed feelings and left the same way, but with greater anxiety. This study shows that there is cause for concern. The Dutch culture does not welcome a critical plurality of opinions regarding the legitimacy of euthanasia. Critics are regarded quite unfavorably.[1] Before I arrived, my review of the contrasting literature led me to think that the Netherlands is neither "promise for the world" or "heaven" nor "hell." My research shows that there is room for concern. Obviously, there is scope for more research, for deeper discussion, and for increased awareness of the pitfalls in the practice of euthanasia.

It was strange for me to discuss the issue of euthanasia in the Netherlands. Views that are extremely unpopular in other countries regarding euthanasia's place in society rule supreme in the Netherlands. The discussions I had with the Dutch experts were almost a mirror image of discussions I had had in Israel, the United States, Britain, Canada and Australia.[2] What was striking in my discussions was the prevailing acceptance of the euthanasia procedure. There were only a few dissenters who were willing to go against the euthanasia policy and practice. My first fourteen interviewees were, on the whole, in favor of the policy, and I felt a growing unease in encountering such unanimity of opinion. This conformity worried me. Plurality and diversity of opinion are good for society, leading to a more comprehensive understanding of the issues, as well as a higher level of truth, as John Stuart Mill used to say.[3]

I found it troublesome that scholars and decision-makers would support a system that suffers from serious flaws when the stakes are very high; after all, we are dealing with life and death.[4] There were various opinions regarding specific questions and issues, but only a minority questioned the system *as such*. Many of the experts depicted a society in which it is the role of doctors to help patients. They did not question the doctors' motives and saw no reason why doctors would perform euthanasia without compelling reasons. They argued that, of course, criminals exist in every society, in every sphere of life, but policy is not built around this small number of criminals. They believed that there is a need to install control mechanisms against the possibility of abuse, but that the system's rationale is good – to help

1 In his comments on the first draft of this study, van Dantzig wrote that this assertion is fundamentally incorrect: "The whole of Dutch society is based on the cohabitation of people who fundamentally disagree on everything. The sometimes very creative solutions (soft drugs may not be bought by coffee shops, but their sale is not punished within certain limits) have given rise to the word "poldermodel," which expressly means living by compromise, or as I have once put it, the fair division of discontent. I write to you because such a fundamental misunderstanding may harm the quality of your paper." Personal communication on July 14, 2000.

2 The outcome of this extensive nine-year research project is *The Right to Die with Dignity: An Argument in Ethics, Medicine and Law* (Piscataway, NJ.: Rutgers University Press, 2001).

3 J. S. Mill, *Utilitarianism, Liberty, and Representative Government* (London: J. M. Dent, 1948), Everyman's edition. For further discussion, see R. Cohen-Almagor, *The Boundaries of Liberty and Tolerance* (Gainesville, FL: The University Press of Florida, 1994); *idem*, "Why Tolerate? Reflections on the Millian Truth Principle," *Philosophia*, Vol. 25, Nos. 1–4 (1997): 131–152; *idem*, "Ends and Means in J.S. Mill's Utilitarian Theory," *The Anglo-American Law Review*, Vol. 26, No. 2 (1997): 141–174.

4 In his comments on the first draft of this essay, Griffiths reacted to this statement by writing: "Nowhere do you suggest that there is a *better* system anywhere else. The Dutch know about the system's defects and are working to improve it. Can you tell me about another country where that is true?" Personal communication on July 10, 2000. Griffiths, it seems, finds a lot of comfort in comparative studies to the point of blurring his own careful thinking about the happenings in his country.

people in their time of need. They emphasized that the two major reports of 1990 and 1995 do not demonstrate a slippery slope, yet ignored the fact that there is already too much abuse.

In this context, James Kennedy made an interesting observation. He said that it is wrong to speak of the euthanasia "system." The word "system" suggests a kind of logical coherence that Kennedy does not think that Dutch euthanasia has. The Dutch use the word "beleid," meaning policy. They are proud of fuzziness. They believe that rules and guidelines should not be set in stone. They do take the Guidelines seriously, but if the physician has a good reason for not fulfilling a certain Guideline, then that is fine. It is hard to know how much weight the Dutch *really* put on their Guidelines. The Guidelines are important but not absolutely important. They carry a certain weight but don't have *absolute* gravity. Kennedy thinks that the new act of April 2001 added a bit more weight to the Guidelines, but only a little bit.[5]

Many of the interviewees failed to recognize that the policy does not work because *all* of the Guidelines, without exception, are broken time and time again.[6] It is not always the patient who makes the request for euthanasia or physician-assisted suicide. Often the doctor proposes euthanasia to the patient. Sometimes the family initiates the request. The requirement that the request be voluntary is thus compromised. On occasion, the patient's request is not well considered. As we have seen, there have been cases in which no request was made and patients were put to death. Furthermore, the patient's request is not always durable and persistent as required.

The Guidelines speak of "unbearable suffering," a term that evokes criticism because it is open to interpretation.[7] Are dementia patients, for instance, suffering unbearably? Were Mrs. Bosscher and Mr. Brongersma in an unbearable state of suffering? The Guidelines instruct that a doctor must perform the euthanasia. Yet, there are cases in which nurses have administered the procedure. It is estimated that 10% of the nursing home physicians let the nurse or even the patient's family members administer the euthanasia drug.[8] Another key Guideline requires that before the doctor assists the patient, a second doctor must be consulted. This Guideline has been breached many times. It is instructed that the patient's GP performs euthanasia, yet on occasion the consultant conducted the final act. The doctor must keep a full written record of every case and report it to the prosecutorial authorities as a case of euthanasia or physician-assisted suicide, and not as a case of death by natural causes. This Guideline has also been violated frequently.[9] Notwithstanding, many interviewees were quite content with the Guidelines. One cannot expect a policy that would work 100 percent of the time, but given the many frequent breaches of all the Guidelines, the Dutch should invest effort to find remedies and to prevent abuse and lawlessness.

5 Interview with James Kennedy on June 24, 2001 (Amsterdam).

6 In his remarks on the first draft of this study, Griffiths wrote that this assertion is "of course pretty silly." He asked: "Do you know of a single legal policy that 'works' 100%? The fact that the Guidelines are not yet effective *enough* does not mean they are having no effect at all. I would argue that the situation in the Netherlands is *much better* than elsewhere, that the difference is that here *we know* the extent to which control is not yet adequate." Personal communication on 10 July 2000.

7 For deliberation on the range of what "unbearable suffering" means, see Carlos F. Gomez, *Regulating Death* (New York: The Free Press, 1991): 99–104.

8 Martien Tom Muller, *Death on Request* (Amsterdam: Vrije Universiteit, 1996), Thesis, p. 52.

9 For deliberation, see Jacqueline M. Cuperus-Bosma, Gerrit van der Wal and Paul J. van der Maas, "Physician-assisted Death: Policy-making by the Assembly of Prosecutors General in the Netherlands," *European J. of Health Law*, Vol. 4 (1997): 225–238.

Upon analyzing the interviewees' answers, we may group them accordingly:

1. Some are avowed advocates of the policy and practice of euthanasia no matter what. They believe that the euthanasia option should be available for patients and are not willing to analyze the situation critically. This group includes A. van Dantzig[10] and Heleen Dupuis.

2. The majority of interviewees defend the practice despite its major flows. One cannot escape noticing that some of them work together in the same institutions and it seems that the working relationships shaped the thinking in certain similar patterns. Some of the interviewees in this group work for government agencies (most notably Visser of the Health Ministry, Leenen who was most instrumental in the legalization of euthanasia, and van der Maas, van der Wal and van Delden who have had major research projects together) and identify with the policy. When the government commissions science, there is always a risk that the scientist will identify with the governmental policy to the point of compromising his or her critical capacity for impartial reflection.[11] Other interviewees in this group are more critically open and think that some accommodations are needed, but that the policy and practice of euthanasia, on the whole, function well. They think that euthanasia should be an option for patients in a liberal society and that, in any event, the Netherlands cannot go back. The public largely supports the policy and wishes it to be continued.[12] This large group consists of Gerrit van der Wal who works at the Free University of Amsterdam; Egbert Schroten and Johannes van Delden who cooperate closely at the Center for Bioethics and Health Law, Utrecht University; Rob Houtepen, Ron Berghmans and George Beusmans from Maastricht; Henk Leenen, emeritus law professor from the University of Amsterdam and J.K. (Sjef) Gevers who succeeded him as the Professor of Health Law at that same university; Jaap Visser of the Ministry of Health; Bert Thijs, a senior physician who heads the VU Hospital Intensive Care Unit[13]; Paul van der Maas and Henri Wijsbek from Erasmus University;[14] Gerrit Kimsma[15] and Bert Keizer who practice euthanasia in the Amsterdam area.

10 In his comments on the first draft, van Dantzig wrote: "Please remove this, this is far from true!" Personal communication on July 8, 2000.

11 In his comments van der Maas reacted to this assertion by saying that he takes it as "very insinuating." Personal communication on September 18, 2000. Leenen wrote in his letter of February 2, 2001 that this comment "is below any level of decent discussion" and that I am going "too far." Leenen maintained that he had not expected me to enter "this type of debate" and that as far as he is concerned "I never did any legislation-work for the government. I have written my law for three parties in the Parliament. The government has taken it over."

12 The number of citizens who approve of euthanasia at the patient's explicit request grew from 40% in 1966 to over 60% (in some polls almost 80%) in 1993. Likewise the number of opponents decreased steadily (21% in 1986, 17% in 1989, 12% in 1994. Cf. Joop van Holsteyn and Margo Trappenburg, "Citizens' Opinions on New Forms of Euthanasia. A Report from the Netherlands," *Patient Education and Counseling*, Vol. 35 (1998), p. 64. A 1998 poll indicated that 92% of the population supports the practice of euthanasia. Cf. "Dutch Might Legalize Euthanasia," *Associated Press* (July 12, 1999).

13 Unlike most interviewees, Professor Thijs did not receive any special training in medical ethics. I asked him: Do senior physicians in your position, i.e., heads of hospital departments, receive any special training in medical ethics? Did you receive any such training? His answer was: "I do not think senior physicians receive special training in medical ethics." Personal communication on December 20, 2000.

14 I asked Paul van der Maas whether he has close working relationships with Henri Wijsbek. His answer was: "I know Dr Wijsbek but we don't have a working relationship." Personal communication on December 19, 2000.

15 In his comments, Kimsma expressed uneasiness about including him in this group. He wrote that the present system is the best or better than any other system. This is the result of deliberations between legal and medical

3. There is a smaller group who recognize the flaws and would like to introduce changes, some of which are quite substantial: John Griffiths, Evert van Leeuwen, Govert den Hartogh, and Arie van der Arend. Like the two former groups, these people still support the practice.

4. The critics of euthanasia, who would like to prohibit the practice entirely, are Frank Koerselman, Chris Rutenfrans and Henk Jochemsen.[16] Jochemsen has strong religious convictions. Chris Rutenfrans is a Roman Catholic who declares that his opposition to euthanasia does not stem from his religion but from practical reasoning.

5. Others recognize that the policy suffers from several serious flaws, some of which may not be correctable. They are struggling with the issue and have ambivalent views about the practice. This group includes Arko Oderwald, Ruud ter Meulen and James Kennedy. Ter Meulen[17] and Kennedy[18] have strong religious convictions.

I was surprised during some of the discussions to hear the rosy pictures that were painted. I asked myself whether I was too cynical and suspicious, or my counterparts too optimistic; after all, they knew the situation in the Netherlands far better than I did. But the unanimity of opinion might suggest that there is not enough reflective thinking about this issue, that the practice of euthanasia is taken for granted;[19] therefore, there might be greater room for abuse in such an environment of trust and lack of critical questioning. Even issues that are acknowledged as

participants of the process over the past 25 years. But it has serious flaws. One of them is the lack of consultation prior to any decision to end a life, where the medical situation and chances for improvement are subject to serious professional deliberations. Kimsma thinks that more medical consultations are necessary before a request for an end to life should be granted, and that "there is a deep confusion about the proper procedures and the proper and overriding frames of thinking that must dictate the procedures. The struggle between law and medicine instead of cooperation has made its victims." Yet, in other countries Kimsma learned from private conversations with physicians that they "had been there" but they didn't talk openly about their practice and didn't consult colleagues. Kimsma concluded that the situation in the Netherlands where "we discuss and practice these issues is the strong part of the present procedure," but much still needs improvement. Other countries need even more improvement. Personal communication on September 9, 2000. I should note that Kimsma and van Leeuwen work very closely together. I assume that the differences of opinion are because Kimsma is a physician who practices euthanasia and van Leeuwen is an ethicist and philosopher.

16 In his comments on the first draft, Jochemsen asked me to add that he does realize that in the present situation a simple reiteration of the prohibition would not improve the practice straight away. This would require a whole package of measures. Personal communication on July 5, 2000.

17 Ruud ter Meulen is the Director of the Institute of Bioethics, and , Maastricht University professor on behalf of the Radboud Foundation, a Catholic organization that sponsors special professorships at non-religious universities. This may explain why his views are so different from the views expressed by his colleagues whose offices are situated next to his at the same university corridor.

18 Kennedy had recently published a book about euthanasia and wished, to the extent he could, to avoid easy prejudice of his own critique of Dutch euthanasia practice. In a personal communication he wrote: "Many Dutch find the religious beliefs of critics as the basis for not considering further what they may have to say. I do intend to refer to my religious faith in my book, but in a way which I hope will enhance, not undermine, the force of my observations. If there is any way in which you can accommodate my concern in your own work, I would appreciate it, though obviously it's your work, and your judgment call." Personal communication dated December 16, 2000.

19 In his comments, Arie van der Arend contested my argument that there is not enough reflective thinking about euthanasia, arguing that (a) I cannot expect extensive and balanced reflective thinking during interviews that were taken from people who were busy with totally different tasks at that moment; (b) my study does not cover the extensive Dutch literature on the subject; (c) I did not interview one of the best "reflective thinkers," Theo Beemer, Professor of Moral Theology and Health Care Ethics at the Catholic University of Nijmegen, and (d) that such a value judgment could have been justified only after comparing the Dutch practice to the situation in other countries. Personal communication on July 3, 2000.

problems are not conceived to be serious enough to press. The Dutch tend to accept highly troublesome contentions and to consider and allow euthanasia in cases where the Guidelines are severly breached and to condone killing when patients did not ask to end their lives. The culture around euthanasia makes the practice accessible within the confines of what is permissible. This culture, however, has a chilling effect upon open, critical debate.[20] In other parts of the world, under similar circumstances and in light of the justified critique, euthanasia would not be considered an option.

B. SUGGESTIONS FOR IMPROVEMENT

Some troubling questions have arisen as a result of this study. The high number of unreported cases of euthanasia is alarming. The fact that some patients have been put to death without prior consent is extremely worrisome. Society has to ensure that no abuse takes place and that the existing legal procedure does not open a window for abuse or provide a way to get rid of "unwanted" patients. More research should be done on what is termed outside of the Netherlands "passive euthanasia", that is, the withholding or withdrawal of treatment. Special attention should be given to demented patients, newborns and children. The Guidelines need to be clarified in detail, closing the door to possible misinterpretation that could lead to abuse.

I agree with most of the experts who contend that euthanasia should not be regarded as an integral part of normal medical care. The fact that many physicians do not wish to be bothered with the procedures is alarming. It shows that they have not internalized the concept that euthanasia is an exceptional medical procedure and, as such, requires social control. It is possible that the moral ambiguity that surrounded the issue—allowing the practice while it was still prohibited under the Penal Code – made doctors feel that they had better conduct euthanasia in private, keeping it only between the patients, their families and themselves. Wherever euthanasia is practiced, it should be subject to constructive criticism. It is preferable to draft a better legal framework than the one that used to govern the practice in the Netherlands prior to the enactment of the new law, which was ambiguous and presented an illegal-yet-tolerated model. In the event that we see that the new euthanasia policy opens the way to abuse, then yet again we should pursue a public debate in which different sectors of society can take part. The understanding of euthanasia and its importance should be changed for it to work properly without abusing the rules of caution.

I also think that physicians should not suggest euthanasia to their patients as an option. By now, the Dutch people are fully aware that euthanasia is available. If patients wish, they can raise the issue themselves. Most of the euthanasia cases involve cancer patients, and at some time during the progressive course of their illness, they can take the initiative and discuss it with their physicians if they are so inclined. If they do not initiate such a discussion, then the physician can assume that the patient does not wish to move in this direction.

I believe that the medical profession should not turn its back on patients who clearly request to shorten their lives. However, this issue should be open to a constant public debate. It is necessary to devise a better working framework to help patients in need. At the same time, respect for human life is and should remain the prime concern. Ending a human life

[20] Hendin reached a similar conclusion. Cf. Herbert Hendin, *Seduced by Death, op. cit.*, p. 100.

without acquiring the patient's consent might be motivated by mercy – or, alternatively, the motivation may be quite different. The ending of a patient's life should be conducted in the light and subject to scrutiny.

As said in *Chapter 3*, before coming to the Netherlands, I supported euthanasia and published some articles calling to recognize the need for euthanasia (in the active sense that it is practiced in the Netherlands). The study in the Netherlands caused me to change my views not on the ethics of euthanasia but rather on the practicality of its implementation. The Dutch study showed that moving from the individual level to the policy level is tricky and complicated. I no longer support euthanasia. Now I restrict my plea to helping patients in need with physician-assisted suicide. This is in order to give patients better control over life and death, until the very last moment, and to provide a further mechanism against abuse. At the same time, I am willing to concede the need for euthanasia in two circumstances: (1) the patient requesting euthanasia is totally paralyzed, from head to toe, and unable to move any muscles that could facilitate assisted suicide; and (2) the patient took oral medication and has been in the process of dying for prolonged hours (see Guideline 11 *infra*).

The majority of Dutch scholars do not share my view. They lump euthanasia and physician-assisted suicide together and even invented an acronym for this purpose: EAS. It should be noted, however, that in August 1995, in an effort to improve the control mechanisms guiding these procedures, the KNMG refined its Guidelines to recommend that assisted suicide, rather than euthanasia, be performed whenever possible.[21]

I believe that the right to die with dignity includes the right to live with dignity until the last minute and the right to part from life in a dignified manner. There are competent, adult patients who feel that the preferable way for them to part from life is through physician-assisted suicide. I propose a circumscribed plea for voluntary physician-assisted suicide on their behalf. The following considerations are pertinent:

Guideline 1. The physician should not suggest assisted suicide to the patient. Instead, it is the patient who should have the option to ask for such assistance. Initiation by the physician might undermine the trust between the patient and the physician, conveying to patients that the doctor is giving up on them and values their lives only to the extent of offering assistance to die. Such an offer might undermine the will to live and to explore further avenues for treatment. Many Dutch physicians do not see this issue as a significant one. Some of them consider it important to raise the issue when it seems that patients do not dare to initiate it on their own. Undoubtedly, however, all people in the Netherlands are aware of the availability of euthanasia and physician-assisted suicide. Any reluctance shown by patients in regard to this issue should be honored and respected.

Guideline 2. The request for physician-assisted suicide of an adult,[22] competent patient who suffers from an intractable, incurable and irreversible disease must be voluntary.[23] The decision is that of the patient who asks to die without pressure, because life appears to be the

21 *Ibid.*, p. 122.

22 Most bill proposals to legislate PAS in the United States specify that the age of the consenting patient must be 18 or older to qualify for the procedure. Russell Korobkin, "Physician-assisted Suicide Legislation: Issues and Preliminary Responses," *Notre Dame J. of Law, Ethics & Public Policy,* Vol. 12, No. 2 (1998), p. 454.

23 See the Dutch requirements of careful practice, in John Griffiths, Alex Bood and Heleen Weyers, *Euthanasia and Law in the Netherlands, op. cit.*, p. 66. See also Chap. II. Article 3 of the Belgian Euthanasia Law adopted by the House of Representatives on May 16, 2002.

worst alternative in the current situation. The patient should state this wish repeatedly over a period of time.[24] We must verify that the request for physician-assisted suicide does not stem from a momentary urge, an impulse, a product of passing depression. This emphasis of enduring request was one of the requirements of the abolished Northern Territory law in Australia,[25] and is one of the requirements of the Oregon *Death with Dignity Act*,[26] as well as of the Dutch Guidelines.[27] We must also verify that the request is not the result of external influences. It should be ascertained with a signed document that the patient is ready to die now, rather than depending solely upon directives from the past. Section 2 of the Oregon Act requires that the written request for medication to end one's life be signed and dated by the patient and witnessed by at least two individuals who, in the presence of the patient, attest to the best of their knowledge and belief that the patient is capable, is acting voluntarily, and is not being coerced to sign the request.[28]

Individuals may express general attitudes regarding euthanasia in an informal discussion made in a social setting, possibly saying that they would not want to live if they were unable to function alone and had to depend on the mercy of others. However, such hypothetical observations do not constitute reliable evidence of a patient's current desires once an actual illness is in progress. This is especially true if the wish was stated when young and healthy. The younger people are and the further they are from serious disease, the more inclined they are to claim that in a hypothetical state of pain, degradation and hopelessness, they would prefer to end their lives. On the other hand, there is a tendency to come to terms with suffering, to compromise with physical disabilities, and to struggle to sustain life. This tendency grows as the body weakens. Many people change their minds when they confront the unattractive alternatives, preferring to remain in what others term the "cruel" world, and continue the Sisyphean struggle for their lives.

Guideline 3. At times, the patient's decision might be influenced by severe pain.[29] In this context, the role of palliative care can be crucial. Ganzini and colleagues report that as a result of palliative care, some patients in Oregon changed their minds about assisted

[24] Many bill proposals to legislate PAS in the United States require a waiting period of 14 or 15 days. Cf. Russell Korobkin, "Physician-assisted Suicide Legislation: Issues and Preliminary Responses," *op. cit.*, p. 468.

[25] Section 7, *Rights of the Terminally Ill Act (1995)* (NT).

[26] In Australia, the law required a "cooling off" period of nine days. In Oregon, the Act requires a waiting period of fifteen days. I do not wish to suggest an arbitrary time period of waiting, saying instead that the patient should state her wish several times "over a period of time." I concur with Miller and colleagues who think that a fifteen days waiting period may be highly burdensome for patients who are suffering intolerably and may preclude access to assisted death for those who request it at the point when they are imminently dying. Franklin G. Miller, Howard Brody and Timothy E. Quill, "Can Physician-Assisted Suicide Be Regulated Effectively?," *J. of Law. Medicine & Ethics*, Vol. 24 (1996), p. 226. See also Oregon *Death with Dignity Act*, *Oregon Revised Statutes*, Vol. 8 (1998 Supplement), at 982.

[27] John Griffiths *et al.*, *Euthanasia and Law in the Netherlands*, *op. cit.*, p. 66.

[28] Oregon *Death with Dignity Act*, *Oregon Revised Statutes*, Vol. 8 (1998 Supplement), at 980.

[29] On the different conceptions of pain that physicians and patients have, see William Ruddick, "Do Doctors Undertreat Pain?" *Bioethics*, Vol. 11, Nos. 3–4 (1997): 246–255. It is argued that pain control represents an often neglected need of nursing home residents, and that nursing home staff underestimate the true pain burden experienced by residents. See Joan M. Teno, Sherry Weitzen, Terrie Wetle and Vincent Mor, "Persistent Pain in Nursing Home Residents," *JAMA*, Vol. 285, No. 16 (April 25, 2001): 2081.

suicide.[30] The World Health Organization defines palliative care as the "active, total care of patients whose disease is not responsive to curative treatment," maintaining that control of pain, of other symptoms, and of psychological, social, and spiritual problems, is paramount.[31] The medical staff must examine whether it is possible to prevent or to ease the pain by means of medication and palliative care.[32] The Oregon *Death with Dignity Act* requires the attending physician to inform the patient of all feasible alternatives, including comfort care, hospice care and pain control.[33] A recent study designed to examine Oregon physicians' attitudes toward and practices regarding care of dying patients since the passage of the Act shows that 30 percent of the 791 respondents reported that they had increased referrals to hospice. Of the 2094 respondents who cared for patients termed "terminally ill," 76 percent reported that they made efforts to improve their knowledge of the use of pain medications.[34] Bill proposals to legislate PAS in Illinois, Hawaii, Maine, Michigan, Vermont, Washington and Wisconsin explicitly require the attending physician to review options for palliative care with the patient, including hospice and/or pain control options.[35] If it is possible to prevent or to ease the patient's pain, then the patient's request for assisted suicide should not be fulfilled; instead, the necessary treatment should be prescribed. This is provided that the educated patient (i.e., a patient who was advised by the medical staff about the available palliative care options) does not refuse to take the painkillers, and that when the pain subsides so does the motive (or one of the main motives) for assisted suicide. If the patient insists on denying all medication, then doctors must first try to find the reasons for this insistence before they comply with the request.[36]

30 Linda Ganzini, Heidi D. Nelson, Terri A. Schmidt, Dale F. Kraemer, Molly A. Delorit, Melinda A. Lee, "Physicians' Experiences with the Oregon Death with Dignity Act," *New Eng. J. of Med.*, Vol. 342, No. 8 (February 24, 2000), p. 563.

31 World Health Organization, *Cancer Pain Relief and Palliative Care: Report of a WHO Expert Committee* (Geneva, Switzerland: World Health Organization, 1990), at 11.

32 Directive 7 in *The General Manager Circular*, Israel Ministry of Health, no. 2/96 (January 31, 1996) holds: "Doctors must concentrate their efforts on easing the pain, torment, and suffering of the patient, a subject of highest priority in medical treatment, especially when terminal patients are concerned," p. 12 (Hebrew). For further deliberation on pain control mechanisms and their importance, see Timothy E. Quill, Bernard Lo, and Dan W. Brock, "Palliative Options of Last Resort," *JAMA*, Vol. 278, No. 23 (December 17, 1997): 2099–2104; P. D. Doyle, G. W. Hanks and N. MacDonald (eds.), *Textbook of Palliative Medicine* (New York: Oxford University Press, 1998); Christine K. Cassel and Kathleen M. Foley, "Principles for Care of Patients at the End of Life: An Emerging Consensus Among the Specialties of Medicine," Milbank Memorial Fund Report (New York, 1999), reported by Vida Foubister, "Medical experts agree on guide for end-of-life care," *American Medical News* (February 7, 2000, http://www.ama-assn.org/amednews/2000/02/07/prsa0207.htm; Timothy E. Quill, Barbara Coombs-Lee and Sally Nunn, "Palliative Treatments of Last Resort: Choosing the Least Harmful Alternative," *Annals of Internal Medicine*, Vol. 132 (March 21, 2000): 488–493. For further discussion on making palliative care decisions for *incompetent* patients, see Jason H.T. Karlawish, Timothy Quill and Diane E. Meier, "A Consensus-Based Approach to Providing Palliative Care to Patients Who Lack Decision-Making Capacity," *Annals of Internal Medicine*, Vol. 130 (May 18, 1999): 835–840.

33 13 Or. Rev. Stat. § 3.01 (1998).

34 Linda Ganzini, Heidi D. Nelson, Melinda A. Lee, Dale F. Kraemer, Terri A. Schmidt and Molly A. Delorit, "Oregon Physicians' Attitudes About and Experiences With End-of-Life Care Since Passage of the Oregon Death with Dignity Act," *JAMA*, Vol. 285 (2001): 2363–2369.

35 Russell Korobkin, "Physician-assisted Suicide Legislation: Issues and Preliminary Responses," *op. cit.*, p. 469.

36 J. V. Lavery, Joseph Boyle, Bernard M. Dickens, Heath Maclean and Peter A. Singer, "Origins of the Desire for Euthanasia and Assisted Suicide in People with HIV-1 or AIDS: A Qualitative Study," *Lancet*, Vol. 358, No. 9279 (August 4, 2001): 362–367. See also Anthony L. Back and Robert A. Pearlman, "Commentary: Desire for Physician-assisted Suicide: Request for a Better Death?," *Lancet*, Vol. 358, No. 9279 (August 4, 2001): 344–345.

At times, coping with pain and suffering can drain all of the patient's emotional strength, exhausting the ability to deal with other issues. In cases of competent patients, the assumption is that the patient understands the meaning of the decision. A psychiatrist's assessment can confirm whether the patient is able to make a decision of such ultimate significance to the patient's life and whether the decision is truly that of the patient, expressed consistently and of his/her own free will. The Northern Territory *Rights of Terminally Ill Act* required that the patient meet with a qualified psychiatrist to confirm that the patient was not clinically depressed.[37] It is worthwhile to hold several such conversations, separated by a few days. The patient's loved ones and the attending physician should be included in at least one of the conversations.

In study findings published in *The Lancet*, Marjolein Bannink and colleagues report on 22 patients (twelve female and ten male) who had psychiatric consultations after requesting immediate euthanasia because of terminal cancer. Ten had the requests granted because psychiatrists determined that they had no psychiatric symptoms and that the desire to die was not a snap decision. Three of them died naturally before their requests could be fulfilled. Twelve other patients were denied euthanasia, either because of significant psychiatric symptoms or because they changed their mind during the evaluation process. In six of the twelve patients, all women with no psychiatric disorders, consultation with a psychiatrist established that the request for euthanasia had not been considered fully. They seemed to regard euthanasia, or having the option of euthanasia, as a way to control their feelings of despair, pain and helplessness. Their appeals were mainly aimed at discussing their situation explicitly with the attending physician, to reduce anxiety and insecurity. Five patients were denied euthanasia because of psychiatric problems. They had cognitive or depressive disorders, which decreased their competence in decision-making. One patient was granted the euthanasia request despite minor psychiatric symptoms. Bannink and colleagues stressed that many cancer specialists may underestimate the psychiatric effects of terminal illness, especially in cases of cancer that has spread to the brain. They urged specialists to consider a psychiatric evaluation if they are not sure.[38]

Guideline 4. The patient must be informed of the situation and the prognosis for recovery or escalation of the disease, with the suffering that it may involve. There must be an exchange

[37] Section 7, *Rights of the Terminally Ill Act (1995)* (NT).

[38] But despite the fact that over half the euthanasia requests were denied or withdrawn after a psychiatrist's evaluation, Bannink and colleagues stopped short of recommending that psychiatrists always be called in on such cases, saying that to do so would inappropriately install them as the moral "gatekeepers" of end-of-life care. Cf. Marjolein Bannink, Arthur R. van Gool, Agnes van der Heide and Paul van der Maas, "Psychiatric Consultation and Quality of Decision Making in Euthanasia," *The Lancet*, Vol. 356 (December 16, 2000), 2067–2068.

of information between doctors and patients.[39] Bearing this in mind, we should be careful to use neutral terms and to refrain from terms that might offend patients and their loved ones.[40]

Guideline 5. It must be ensured that the patient's decision is not a result of familial and environmental pressures. At times, patients may feel that they constitute a burden on their loved ones. It is the task of social workers to examine patients' motives and to see to what extent they are affected by various external pressures (as opposed to a truly free will to die). A situation could exist in which the patient is under no such pressure, but still does not wish to be a burden on others. Obviously, we cannot say that the feelings of patients toward their loved ones are not relevant to the decision-making process.

Guideline 6. The decision-making process should include a second opinion in order to verify the diagnosis and minimize the chances of misdiagnosis, as well as to allow the discovery of other medical options. A specialist, who is not dependent on the first doctor, either professionally or otherwise, should provide the second opinion.[41] The patient's attending physician, who supposedly knows the patient's case better than any other expert, must be consulted, and all reasonable alternative treatments must be explored. The Oregon *Death with Dignity Act* requires that a consulting physician shall examine the patient and his/her relevant medical records and subsequently confirm, in writing, the attending physician's diagnosis that "the patient is suffering from a terminal disease." Furthermore, the consulting physician must verify that the patient is capable, is acting voluntarily, and has made an informed decision.[42] The Dutch Guidelines require that the physician consult a colleague.[43] The Northern Territory *Rights of Terminally Ill Act* required that a physician who specialized in treating terminal illness examine the patient.[44]

[39] On this issue, see Oregon *Death with Dignity Act,* Section 3, Attending physician responsibilities. Many bill proposals to legislate PAS in the United States specify certain information that must be communicated by the physician to the patient before honoring her request. Cf. Russell Korobkin, "Physician-assisted Suicide Legislation: Issues and Preliminary Responses," *op. cit.,* p. 468. See also See also Chap. II, Article 3 of the Belgian Euthanasia Law, and Section D: Consent to Medical Treatment of The Israel *Patients' Rights Law*, 1992, Law Proposal 2132 (March 16, 1992); The *Patients' Rights Law*, 1996, *Israel Book of Laws*, 1591 (May 12, 1996), at 329–331; and *The General Manager Circular*, The Ministry of Health, no. 2/96 (January 31, 1996), at 10–11 (Hebrew).

[40] R. Cohen-Almagor, "Language and Reality at the End of Life", *Journal of Law, Medicine and Ethics*, Vol. 28, No. 3 (Fall 2000): 267–278.

[41] Most bill proposals to legislate PAS in the United States required that the treating physician refer the patient to a second consulting physician to verify the terminal nature of the disease. The Massachusetts bill required a third confirming opinion. Cf. Russell Korobkin, "Physician-assisted Suicide Legislation: Issues and Preliminary Responses," *op. cit.*, p. 453. See also Chap. II, Article 3 of the Belgian Euthanasia Law.

[42] Oregon *Death with Dignity Act, Oregon Revised Statutes,* Vol. 8 (1998 Supplement), at 981–982. The American Medical Association's council on ethical and judicial affairs suggests the participation of consultants to facilitate discussions that would help the parties reach a course of action. AMA strongly objects to PAS. See Fred Charatan, "AMA Issues Guidelines on End of Life Care," *BMJ*, Vol. 318 (March 13, 1999): 690.

[43] John Griffiths *et al., Euthanasia and Law in the Netherlands, op. cit.*, pp. 66, 104. The Dutch Guidelines require the doctor to consult an independent colleague, not in order to advise the first doctor on medical treatment, but in order to verify whether the criteria of the Guidelines have been satisfied. The consultation is about the patient's condition and life expectancy, the available alternatives and the adequacy of the request. Medical consultation in an earlier stage is part of normal practice. For instance, cancer patients who request euthanasia have invariably been treated in hospitals up to the point at which the doctors and the patient together decided to stop treatment.

[44] Section 7, *Rights of the Terminally Ill Act (1995)* (NT).

Guideline 7. It is advisable for the identity of the consultant to be determined by a small committee of specialists (like SCEN), who will review the requests for physician-assisted suicide. This is in order to avoid the possibility of arranging deals between doctors ("you will consult for me regarding Mr. Jones, approving my decision, and I will consult for you regarding Ms. Smith, approving your decision").

Guideline 8. Some time prior to the performance of physician-assisted suicide, a doctor and a psychiatrist are required to visit and examine the patient so as to verify that this is the genuine wish of a person of sound mind who is not being coerced or influenced by a third party. The conversation between the doctors and the patient should be held without the presence of family members in the room in order to avoid familial pressure. A date for the procedure is then agreed upon.[45] The patient's loved ones will be notified so that they can be present right until the performance of the act, making the day an intimate, family occasion.

Guideline 9. The patient can rescind at any time and in any manner. This provision was granted under the Australian Northern Territory Act[46] and under the Oregon *Death with Dignity Act.*[47] Chapter III, Article 4 of the Belgian Euthanasia Law says that patients can withdraw or adjust their euthanasia declaration at any time.[48]

Guideline 10. Physician-assisted suicide may be performed only by a doctor and in the presence of another doctor. The decision-making team should include at least two doctors and a lawyer, who will examine the legal aspects involved. Insisting on this protocol would serve as a safety valve against possible abuse. Perhaps a public representative should also be present during the entire procedure, including the decision-making process and the performance of the act. This extra caution should ensure that the right to die with dignity does not become a duty. The doctor performing the assisted suicide should be the one who knows the patient best, having been involved in the patient's treatment, taken part in the consultations, and verified through the help of social workers, nurses and psychologists that euthanasia is the true wish of the patient.

Guideline 11. Physician-assisted suicide may be conducted in one of three ways, all of them discussed openly and decided upon by the physician and the patient together: (1) oral medication; (2) self-administered, lethal intravenous infusion; (3) self-administered lethal injection. Oral medication may be difficult or impossible for many patients to ingest because of nausea or other side effects of their illnesses. In the event that oral medication is provided

45 Proposals to legislate PAS in Illinois, Massachusetts and Maine required that a patient seeking PAS obtain a consultation with a mental health professional in order to insure that the patient can pass the "impaired judgment" standard. Cf. Russell Korobkin, "Physician-assisted Suicide Legislation: Issues and Preliminary Responses," *op. cit.*, p. 456. This Guideline is somewhat similar to the guidelines of the Swiss EXIT protocol. See South Australian Voluntary Euthanasia Society, DID YOU KNOW? Assisted Suicide in Switzerland – SAVES Fact Sheet No. 20, issued February 1997. Correspondence with: Hon. Secretary, SAVES, PO Box 2151, Kent Town, SA 5071, Australia – Fax + 61 8 8265 2287. URL: http://www.finalexit.org/

46 Andrew L. Plattner, "Australia's Northern Territory: The First Jurisdiction to Legislate Voluntary Euthanasia, and the First to Repeal It," *DePaul J. of Health Care Law,* Vol. 1 (spring 1997), p. 648.

47 13 Or. Rev. Stat. § 3.07 (1998).

48 Chamber of the Representatives of Belgium, *Government Bill Related to Euthanasia* (November 5, 2001), Doc 5014488/001, project transmitted by the Senate.

and the dying process is lingering on for long hours, the physician is allowed to administer a lethal injection.[49]

Guideline 12. Doctors may not demand a *special fee* for the performance of assisted suicide. The motive for physician-assisted suicide is humane, so there must be no financial incentive and no special payment that might cause commercialization and promotion of such procedures.

Guideline 13. There must be extensive documentation in the patient's medical file, including the following: diagnosis and prognosis of the disease by the attending and the consulting physicians; attempted treatments; the patient's reasons for seeking physician-assisted suicide; the patient's request in writing or documented on a video recording; documentation of conversations with the patient; the physician's offer to the patient to rescind her request; documentation of discussions with the patient's loved ones; and a psychological report confirming the patient's condition. This meticulous documentation is meant to prevent exploitation of any kind -- personal, medical, or institutional.[50] Each report should be examined by a coroner following completion of the physician-assisted suicide.[51]

Guideline 14. Pharmacists should also be required to report all prescriptions for lethal medication, thus providing a further check on physicians' reporting.

Guideline 15. Doctors must not be coerced into taking actions that contradict their conscience or their understanding of their role. This was provided under the Northern Territory Act.[52]

Guideline 16. The local medical association should establish a committee, whose role will be not only to investigate the underlying facts that were reported but also to investigate whether there are "mercy" cases that were not reported and/or that did not comply with the Guidelines.

Guideline 17. Licensing sanctions will be taken to punish those health care professionals who violated the Guidelines, failed to consult or to file reports, engaged in involuntary

49 Cf. Johanna H. Groenewoud, Agnes van der Heide, Bregje D. Onwuteaka-Philipsen *et al,* "Clinical Problems with the Performance of Euthanasia and Physician-assisted Suicide in the Netherlands," *N. Eng. J. Med.*, Vol. 342, No. 8 (2000): 551–556.

50 For further deliberation, see the Dutch Guidelines in John Griffiths *et al., Euthanasia and Law in the Netherlands, op. cit.,* p. 66; Chap. II, Article 3 of the Belgian Euthanasia Law; Oregon *Death with Dignity Act, Oregon Revised Statutes,* Vol. 8 (1998 Supplement), Section 3, at 983. Rebecca Cook pointed out in her comments that such a bureaucratic procedure might discriminate against minorities who will not find it easy to cope with the described demands. However, the demand for detailed documentation is meant to prevent abuse. We should be sensitive to cultural differences and strive to meet special needs that arise from cultural norms but not at the expense of opening the door wide for 'eliminating' unwanted people.

51 Directive 6 in *The General Manager Circular*, Israel Ministry of Health, no. 2/96 (January 31, 1996) states: "The decision to respect a patient's objection to a life prolonging treatment shall be documented in the medical statutes, expressing maximum reasons for the decision and the discussions with the patient," p. 12 (Hebrew). See also Israel *Patients' Rights Law* (1996), 1591, Chapter E: medical documentation and medical information, p. 331.

52 Andrew L. Plattner, "Australia's Northern Territory: The First Jurisdiction to Legislate Voluntary Euthanasia, and the First to Repeal It," *op. cit.,* p. 648. The Illinois proposed bill to legislate PAS included a "Provider's Freedom of Conscience" clause, which explicitly said that physicians who object to the practice may not be required to participate or aid in PAS. Cf. Russell Korobkin, "Physician-assisted Suicide Legislation: Issues and Preliminary Responses," *op. cit.,* p. 464.

euthanasia without the patient's consent or with patients lacking proper decision-making capacity. Physicians who failed to comply with the above Guidelines will be charged and procedures to sanction them will be brought by the Disciplinary Tribunal of the Medical Association. The maximum penalty for violation of the Guidelines will be the revoking of the physician's medical license. In the event that this penalty proves insufficient in deterring potential abusers, there will be room to consider further penalties, including heavy fines and prison sentences.[53]

What is presented here is a circumscribed reasoning for physician-assisted suicide to help a designated group of patients that require and deserve help from the medical profession in departing from life. With human life at stake, detailed procedures are required in order to prevent abuse. I suggest that this reasoning be adopted for a trial period of one year, at which time the consequences will be examined to determine whether further implementation of the policy for a lengthy period of time is justified. During this one-year trial period, feedback between physicians, ethicists and the public at large in reviewing the policy and practice of physician-assisted suicide should be welcomed and encouraged. If the proposal fails (for instance, if physicians do not adequately report incidents of physician-assisted suicide), then all the data should be brought before a review committee to closely study the policy and practice. Members of the committee will issue a report recommending whether they wish to continue the practice, to amend the Guidelines, or to abolish physician-assisted suicide entirely. Preferably, the final decision should be made through active participation of all those involved.

[53] For further deliberation, see Arthur L. Caplan, Lois Snyder and Kathy Faber-Langendoen, "The Role of Guidelines in the Practice of Physician-Assisted Suicide," *Annals of Internal Medicine,* Vol. 132 (March 21, 2000): 476–481; S. Frileux, C. Lelièvre, M. T. Muñoz Sastre, E. Mullet, and P. C. Sorum, "When Is Physician Assisted Suicide or Euthanasia Acceptable?," *Journal of Medical Ethics,* Vol. 29, Issue 6 (December 1, 2003): 330-336.

APPENDIX I:

INTERVIEWS IN THE NETHERLANDS (SUMMER 1999)

Professor John Griffiths, Department of Legal Theory, Faculty of Law, University of Groningen (Groningen, July 16, 1999).

Professor J.K. Gevers, Professor of Health Law, University of Amsterdam (Amsterdam, July 19, 1999).

Professor Evert van Leeuwen, Department of Metamedicine, Free University of Amsterdam (Amsterdam, July 19, 1999; Haarlem, July 28, 1999).

Professor Bert Thijs, Medical Intensive Care Unit, VU Hospital, Amsterdam (Amsterdam, July 20, 1999).

Professor A. van Dantzig, retired expert in psychiatry (Amsterdam, July 20, 1999).

Professor H.J.J. Leenen, formerly professor of social medicine and health law, Medical Faculty and Faculty of Law, University of Amsterdam (Amsterdam, July 21, 1999).

Professor Gerrit van der Wal, Institute for Research in Extramural Medicine, Department of Social Medicine, Free University of Amsterdam (Amsterdam, July 21, 1999).

Dr. Jaap J.F. Visser, Ministry of Health, Department of Medical Ethics, The Hague (Amsterdam, July 21, 1999).

Professor Heleen Dupuis, Department of Metamedicine, University of Leiden (Leiden, July 22, 1999).

Dr. Henri Wijsbek, Department of Medical Ethics, Erasmus University of Rotterdam (Rotterdam, July 23, 1999).

Dr. Arie J.G. van der Arend, Health Ethics and Philosophy, Maastricht University (Maastricht, July 26, 1999).

Dr. George Beusmans, Maastricht Hospital (Maastricht, July 26, 1999).

Professor G.F. Koerselman, Sint Lucas Andreas Hospital, Amsterdam (Amsterdam, July 27, 1999).

Professor Henk Jochemsen, Professor Lindeboom Institute (Ede Waginingen, July 27, 1999).

Dr. Gerrit K. Kimsma, Department of Metamedicine, Free University of Amsterdam (Koog aan de Zaan, July 28, 1999).

Dr. James Kennedy, Department of History, Hope College, Michigan. Visiting Research Fellow at the Institute for Social Research, Amsterdam (Amsterdam, July 29, 1999).

Professor Paul van der Maas, Director, Department of Public Health, Faculty of Medicine, Erasmus University, Rotterdam (Amsterdam, July 29, 1999).

Dr. Chris Rutenfrans, *Trouw* (Amsterdam, July 30, 1999).

Dr. Arko Oderwald, Department of Metamedicine, Free University of Amsterdam (Amsterdam, July 30, 1999; August 8, 1999).

Ms. B. de B. and her three children (Amsterdam, August 2, 1999).

Professor Egbert Schroten, Center for Bioethics and Health Law, Utrecht University (Utrecht, August 5, 1999).

Professor Govert den Hartogh, Faculty of Philosophy, University of Amsterdam (Amsterdam, August 10, 1999).

Dr. Johannes JM van Delden, Senior Researcher, Center for Bioethics and Health Law, Utrecht University (Utrecht, August 10, 1999).

Dr. Rob Houtepen, Health Ethics and Philosophy, Maastricht University (Maastricht, August 11, 1999).

Dr. Ron Berghmans, Institute for Bioethics, Maastricht University (Maastricht, August 11, 1999).

Professor Ruud ter Meulen, Director, Institute for Bioethics and Professor at the University of Maastricht (Maastricht, August 11, 1999).

APPENDIX II

INTERVIEWS AND TELEPHONE CONVERSATIONS IN THE NETHERLANDS (SUMMER 2001)*

Dr. Jaap J.F. Visser, Ministry of Health, Department of Medical Ethics, The Hague (phone conversation on June 20, 2001).
Professor Evert van Leeuwen, Department of Metamedicine, Free University of Amsterdam (Amsterdam, June 20, 2001).
Professor A. van Dantzig, retired expert in psychiatry (phone conversation on June 21, 2001).
Dr. Gerrit K. Kimsma, Department of Metamedicine, Free University of Amsterdam (Amsterdam, June 21, 2001).
Professor Ruud ter Meulen, Director, Institute for Bioethics and Professor at the University of Maastricht (phone conversation on June 22, 2001).
Dr. Chris Rutenfrans, *Trouw* (Amsterdam, June 22, 2001).
Dr. James Kennedy, Department of History, Hope College, Michigan (Amsterdam, June 24, 2001).
Professor J.K. Gevers, Professor of Health Law, University of Amsterdam (phone conversation on June 25, 2001).
Dr. Arko Oderwald, Department of Metamedicine, Free University of Amsterdam (Amsterdam, June 25, 2001).

* Seven interviewees answered my queries by writing: Frank Koerselman, John Griffiths, H.J.J. Leenen, Arie J.G. van der Arend, Henk Jochemsen, Rob Houtepen and Ron Berghmans. In addition, Govert den Hartogh and Paul van der Maas provided a partial response.

APPENDIX III

INTERVIEWS IN THE NETHERLANDS (APRIL 2002)

Dr. Arie J.G. van der Arend, Health Ethics and Philosophy, Maastricht University (Maastricht, April 2, 2002).

Dr. Rob Houtepen, Health Ethics and Philosophy, Maastricht University (Maastricht, April 2, 2002).

Professor Ruud ter Meulen, Director, Institute for Bioethics and Professor at the University of Maastricht (Maastricht, April 2, 2002).

Dr. Ron Berghmans, Institute for Bioethics, Maastricht University (Maastricht, April 2, 2002).

Dr. Jaap J.F. Visser, Ministry of Health, Department of Medical Ethics (The Hague, April 3, 2002).

Dr. Arko Oderwald, Department of Metamedicine, Free University of Amsterdam (Amsterdam, April 3, 2002).

Dr. Bert Keizer, Vreugdehof Nursing Home (Amsterdam, April 3, 2002).

Professor Henk Jochemsen, Free University (Amsterdam, April 4, 2002).

Professor G.F. Koerselman, Sint Lucas Andreas Hospital, Amsterdam (Amsterdam, April 4, 2002).

Professor Evert van Leeuwen, Department of Metamedicine, Free University of Amsterdam (Amsterdam, April 8, 2002).

INDEX - GENERAL

INDEX – COURT CASES

International Library of Ethics, Law, and the New Medicine

1. L. Nordenfelt: *Action, Ability and Health.* Essays in the Philosophy of Action and Welfare. 2000 ISBN 0-7923-6206-3
2. J. Bergsma and D.C. Thomasma: *Autonomy and Clinical Medicine.* Renewing the Health Professional Relation with the Patient. 2000 ISBN 0-7923-6207-1
3. S. Rinken: *The AIDS Crisis and the Modern Self.* Biographical Self-Construction in the Awareness of Finitude. 2000 ISBN 0-7923-6371-X
4. M. Verweij: *Preventive Medicine Between Obligation and Aspiration.* 2000 ISBN 0-7923-6691-3
5. F. Svenaeus: *The Hermeneutics of Medicine and the Phenomenology of Health.* Steps Towards a Philosophy of Medical Practice. 2001 ISBN 0-7923-6757-X
6. D.M. Vukadinovich and S.L. Krinsky: *Ethics and Law in Modern Medicine.* Hypothetical Case Studies. 2001 ISBN 1-4020-0088-X
7. D.C. Thomasma, D.N. Weisstub and C. Hervé (eds.): *Personhood and Health Care.* 2001 ISBN 1-4020-0098-7
8. H. ten Have and B. Gordijn (eds.): *Bioethics in a European Perspective.* 2001 ISBN 1-4020-0126-6
9. P.-A. Tengland: *Mental Health.* A Philosophical Analysis. 2001 ISBN 1-4020-0179-7
10. D.N. Weisstub, D.C. Thomasma, S. Gauthier and G.F. Tomossy (eds.) : *Aging: Culture, Health, and Social Change.* 2001 ISBN 1-4020-0180-0
11. D.N. Weisstub, D.C. Thomasma, S. Gauthier and G.F. Tomossy (eds.) : *Aging: Caring for our Elders.* 2001 ISBN 1-4020-0181-9
12. D.N. Weisstub, D.C. Thomasma, S. Gauthier and G.F. Tomossy (eds.) : *Aging: Decisions at the End of Life.* 2001 ISBN 1-4020-0182-7
(Set ISBN for Vols. 10-12: 1-4020-0183-5)
13. M.J. Commers: *Determinants of Health: Theory, Understanding, Portrayal, Policy.* 2002 ISBN 1-4020-0809-0
14. I.N. Olver: *Is Death Ever Preferable to Life?* 2002 ISBN 1-4020-1029-X
15. C. Kopp: *The New Era of AIDS.* HIV and Medicine in Times of Transition. 2003 ISBN 1-4020-1048-6
16. R.L. Sturman: *Six Lives in Jerusalem.* End-of-Life Decisions in Jerusalem - Cultural, Medical, Ethical and Legal Considerations. 2003 ISBN 1-4020-1725-1
17. D.C. Wertz and J.C. Fletcher: *Genetics and Ethics in Global Perspective.* 2004 ISBN 1-4020-1768-5
18. J.B.R. Gaie: *The Ethics of Medical Involvement in Capital Punishment.* A Philosophical Discussion. 2004 ISBN 1-4020-1764-2
19. M. Boylan (ed.): *Public Health Policy and Ethics.* 2004 ISBN 1-4020-1762-6; Pb 1-4020-1763-4
20. R. Cohen-Almagor: *Euthanasia in the Netherlands.* The Policy and Practice of Mercy Killing. 2004 ISBN 1-4020-2250-6

KLUWER ACADEMIC PUBLISHERS – DORDRECHT / BOSTON / LONDON

MIX
Papier aus verantwortungsvollen Quellen
Paper from responsible sources
FSC® C105338

If you have any concerns about our products,
you can contact us on
ProductSafety@springernature.com

In case Publisher is established outside the EU,
the EU authorized representative is:
Springer Nature Customer Service Center GmbH
Europaplatz 3, 69115 Heidelberg, Germany

Printed by Libri Plureos GmbH
in Hamburg, Germany